STRUCTURAL AND MORPHOLOGICAL EVOLUTION IN METAL–ORGANIC FILMS AND MULTILAYERS

STRUCTURAL AND MORPHOLOGICAL EVOLUTION IN METAL–ORGANIC FILMS AND MULTILAYERS

ALOKMAY DATTA

SAHA INSTITUTE OF NUCLEAR PHYSICS, KOLKATA, INDIA

SMITA MUKHERJEE

SAHA INSTITUTE OF NUCLEAR PHYSICS, KOLKATA, INDIA

CRC Press
Taylor & Francis Group
Boca Raton London New York

CRC Press is an imprint of the
Taylor & Francis Group, an **informa** business

CRC Press
Taylor & Francis Group
6000 Broken Sound Parkway NW, Suite 300
Boca Raton, FL 33487-2742

First issued in paperback 2019

© 2016 by Taylor & Francis Group, LLC
CRC Press is an imprint of Taylor & Francis Group, an Informa business

No claim to original U.S. Government works

ISBN-13: 978-1-4822-3270-7 (hbk)
ISBN-13: 978-0-367-37737-3 (pbk)

Visit the Taylor & Francis Web site at
http://www.taylorandfrancis.com

and the CRC Press Web site at
http://www.crcpress.com

Contents

Foreword

Two female scientists tower over the field of organic monolayers and multilayers. In 1891, *Nature* published a remarkable paper titled Surface Tension. The manuscript had been sent to Lord Rayleigh from Germany by Agnes Pockels, who did not have (and at that time could not have obtained) a formal scientific education, but had performed groundbreaking experiments on floating monolayers in her kitchen. Some decades later, in 1918, Katherine Blodgett received a master's degree from the University of Chicago and became the first woman ever hired by the General Electric research laboratory. Working with Irving Langmuir, perhaps the field's most well-known figure, she developed the Langmuir–Blodgett method for building up organic multilayers using layer-by-layer deposition of floating monolayers. It is important to realize that, long before words such as surface science and nanotechnology came into vogue, these scientists were manipulating nanoscale layers of molecules at surfaces, and building artificially structured materials with designed properties and practical applications. (Reportedly, the movie *Gone with the Wind* used Langmuir–Blodgett multilayers as antireflective coatings for their camera lenses.) And they were doing all this not with millions of dollars of shiny stainless steel equipment but with basic, often store-bought tools. Because of the ease and power of these techniques, they have never really gone out of use, and the subject has enjoyed a renaissance in recent years. This revival has been driven by new scientific questions and new molecular materials, but perhaps most of all it has been transformed by new characterization techniques that Pockels, Langmuir, and Blodgett could never have imagined. My own expertise is in the area of x-ray scattering, but methods such as ellipsometry, x-ray fine structure spectroscopy (XAFS), Fourier transform infrared (FTIR) spectroscopy and atomic force microscopy have combined with x-ray reflectivity and grazing incidence diffraction to give us a remarkably detailed view of how the molecules and ions are arranged in molecular thin films as well as of their growth processes and dynamics. In 1966 George Gaines published *Insoluble Monolayers at the Air–Water Interface*, which soon became and long remained the indispensable reference book for the field. Smita Mukherjee and Alokmay Datta of the Saha Institute of Nuclear Physics, one of the world's leading centers for monolayer and multilayer studies, have now written a worthy successor to Gaines's book. It is interesting to compare the two books and see how much has changed. The scientific world has come a long way in these fifty years, and this book describes in remarkable detail where we are today and where we are going. Of course there is still much left to do. Those who wish to learn what has been done, or who aspire to do more, would be wise to start by reading this book.

Pulak Dutta
Northwestern University, Evanston, Illinois, USA

Preface

This book presents the major results of the work we carried out on Langmuir monolayers and Langmuir–Blodgett multilayers from 2004 to 2012, as the doctoral research of one of us (Smita) under the supervision of the other (Alokmay), and has been updated to the best of our ability. The last phrase is important. Though organic films have been spread on water and transferred onto substrates since Hammurabi's time, i.e., more than 4000 years ago and has been scientifically studied for more than 130 years (beginning with Agnes Pockels), yet the enigma of such a simply achieved two-dimensional system still holds physicists, chemists, and biologists enthralled and the field continuously produces new insights and also new puzzles leading to new research.

We were fascinated and focused on two aspects of these systems. The first concerns the nature of these two-dimensional objects: Are they more like "solids" or like "liquids?" For quite a long time they have been placed under the category of "two-dimensional liquid crystals" and their phases have been named in accordance with that classification. These include pure monolayers such as those constituting amphiphilic fatty acids. However, during the last two decades, it has been shown that the presence of metal ions can play havoc with this pigeon-holing. Also, it was in general not clear what constitutes a "solid" or "liquid" in two dimensions until the end of the 1980s. Fortunately now we have new ways of probing the structures of two-dimensional things, which have expanded and generalized our notions of solids and liquids in such circumstances and we have addressed this question in these metal–organic monolayer systems armed with these tools.

The other question which intrigued and motivated us was the effect of the lowering of dimensionality on bonding. To put it in a different way: Does a two-dimensional system have different kinds of bonds than a bulk or three-dimensional body? The answer to that, we felt, would tell something about the "weird" chemistry these materials have. As any chemist will tell you, it is impossible to prepare transition element salts of long-chain fatty acids by direct reaction of the fatty acid with the oxide or hydroxide of the transition element. In contrast, as early as the 1930s Irving Langmuir showed that a monolayer of fatty acid on a water surface could "sense" and react to 3×10^{-10} g cm^{-3} of Al ions, or around 7 Al ions per μm^{-3} of water! Hence if you want to prepare, say, cobalt stearate all you need to do is spread a monolayer of amphiphilic stearic acid on water containing cobalt ions as obtained by dissolving cobalt chloride, and the cobalt ions will do the rest themselves. If you now transfer the monolayer onto any substrate and go on repeating the process (as Katherine Blodgett did) you end up with a multilayer of cobalt stearate with some unreacted acid that can be easily washed away. This made us wonder whether the acid molecule at the water surface is the same as that in the bulk body and we found that in fact, it is not.

When, in the course of our work, we got some answers to our queries and published the results in journals, we were lucky to have two sources of inspiration in the form of Professor Gour Prasad Das of the Indian Association for the Cultivation of Science, a premier institute in India and Aastha Sharma of Taylor & Francis. While the first ignited the spark of an idea about writing a book that covers these newfound aspects of these two-dimensional, complex, soft materials, the latter with immense patience has kept the flame burning. This book owes its existence to these two, especially the latter. Then there is Professor Pulak Dutta of Northwestern University, USA, who over and above being the guiding light of one of us (Alokmay) in his research on Langmuir monolayers, agreed to write the foreword for the book, to our great delight and thankfulness. Uttam Basak, our friend (and Alokmay's student), took the trouble of writing the entire section on ellipsometry and Brewster angle microscopy for us.

We should humbly submit that this book is far from an exhaustive survey. We were interested in giving our readers glimpses of the richness hidden in these supposedly well-known systems and, among the amphiphiles, we chose the most well-known, the saturated, aliphatic fatty acids. They have been the "physicists' workhorses," and we also started looking for the mechanisms behind the growth of monolayers and multilayers of these molecules primarily from a physicist's point of view. However, when we brought in the metal ions we saw some interplay between physics and chemistry. We hope that, if nothing else, this book may lead people to take more interest in these issues and ask more questions. In fact, that will be much more than we can expect.

There are a number of people who stood by us in our research. Professor Milan Kumar Sanyal was instrumental in bringing research on surfaces and interfaces to our institute and building up our whole group. Without his active leadership work on monolayers, multilayers and nanomaterials could not have taken the shape we have given them. Professor Stefano Nannarone of the University of Modena e Reggio Emilia and in charge of the BEAR beamline at ELETTRA Synchrotron, Italy, and his dedicated beamline scientists, Dr. Angelo Giglia and Dr. Nicola Mahne, helped us immensely at the beamline and were ready (even at 2 a.m.!) to discuss the nuances of XAFS measurements in the context of our samples. In this connection we gratefully acknowledge the financial support of the Department of Science and Technology, India, and the Programme of Cooperation, Italy, which made these experimental visits possible.

Both of us are extremely fortunate to have immensely supportive families. Our spouses Biswajit and Neepa have been our fountains of strength throughout. Little Bitan (Smita's son), with his happy face and Swapnasopan (Alokmay's son), with his fresh and incisive questions, have provided the ballast to move on with our work. Finally, we can never forget the influences of our parents in shaping our paths, both in and out of the lab.

Chapter 1

Monolayers and Multilayers

1.1 About This Book

Ultrathin and thin films with carefully designed structures and properties have attracted a great deal of interest in recent years [1–5] due to their potential applications in a number of different fields, such as sensors [6–8], detectors [1–3], surface coating [4,5], optical signal processing [9–11], digital optical switching devices [12–14], molecular electronic devices [15–21], nonlinear optics and models mimicking biological membranes [22–24]. These applications require, in general, well ordered films consisting of molecules with specific properties, carefully aligned with respect to each other and to the substrates, and possessing high degree of stability to thermal and chemical changes. Langmuir–Blodgett (LB) technique is a simple but powerful tool for creating carefully controlled supramolecular structures of organized molecular assemblies. The structure, morphology, configuration and various other physical properties of the films obtained by LB technique can be easily modified to suit any specific application. The possibility to synthesize organic molecules almost without limitations and with desired structure and functionality, in conjunction with Langmuir–Blodgett film deposition technique, enables the production of electrically, optically, and biologically active components on a nanometer scale. It is thus extremely important to get an idea about the relation between the molecular structure and the domain structure on one hand and the various physical properties of such systems on the other. The basic physics involved in such interrelations, as we shall see in this book, is a topic of fundamental importance.

This book relates some structural, bonding and morphological aspects of ultrathin metal–organic films deposited by the LB technique. Before proceeding on to discuss the outline of our work, in this introductory chapter, we have talked about Langmuir monolayers and Langmuir–Blodgett films, basics of their growth mechanism and some related aspects of metal–organic complexes, along with a brief review of initial scientific observations.

1.2 Langmuir Monolayers

In 1917, Irving Langmuir developed experimental and theoretical concepts which underlie present day understanding of behavior of molecules in insoluble monolayers on water [25]. Organic films can be formed on the water surface very easily by choosing a special type of molecule called an *amphiphilic molecule* [3, 26, 27]. These molecules have two parts, one part (called head) is soluble in water, i.e., hydrophilic and the other part (called tail) is insoluble in water, i.e., hydrophobic (Fig. 1.1). When such molecules are spread on water, the amphiphilic molecules produce a monomolecular layer with the heads touching water and the tails pointing

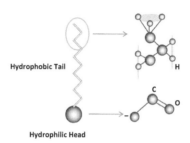

Figure 1.1: *Long chain fatty acid: A typical amphiphile.*

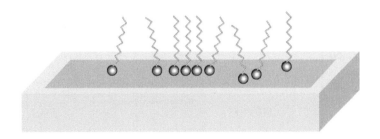

Figure 1.2: *Langmuir monolayer.*

towards air (Fig. 1.2). The monomolecular layer thus formed is generally called a *Langmuir monolayer (LM)* [3,26,27]. Long chain fatty acids, alcohols, lipids, etc., are the most common examples of such molecules. These molecules, which are also commonly called surfactants, can form stable monolayers depending upon their hydrophilicity of headgroups and hydrophobicity of tails.

1.2.1 Structure and Properties

There are several properties of the monolayer, like surface pressure, surface potential [5, 26, 28] and surface viscosity [3, 5, 26]; the most important being its surface tension. However, the physically measurable quantity is the surface pressure [25], i.e., the difference between the surface tension of pure water surface and that covered by the film. Surface pressure π_s [5] is thus defined as

$$\pi_s = \gamma_w - \gamma \tag{1.1}$$

where γ_w is the surface tension of pure water and γ is the surface tension of the film covered surface. The variation of the surface pressure with surface area [29], temperature [27, 29], metal ions [30, 31], subphase pH [32], compression rate [33], time [34] have been extensively studied. The most important and basic information about a monolayer is its surface pressure (π_s) - specific molecular area (A) isotherm which is the two-dimensional equivalent of the p (pressure)- V (volume) isotherm for a bulk system. Fig. 1.3 shows the schematic of a typical surface pressure (π_s)-specific molecular area (A) isotherm for a Langmuir monolayer.

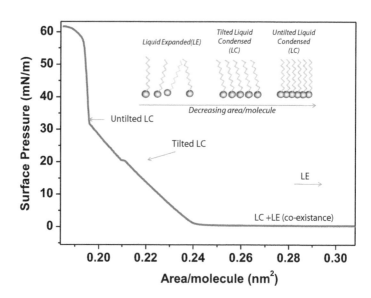

Figure 1.3: *Typical surface pressure versus specific molecular area isotherm for behenic acid. Flat part and kink in the isotherm are due to phase transitions.*

Langmuir monolayers are used as models for two-dimensional systems, as they are easily achieved and are robust enough for studying structural changes [27] and phase transitions in two-dimensions [35, 36]. Other than this basic interest, Langmuir monolayers are used of course as precursors for LB films on solid substrates [37, 38] and also can be used as templates for bio-mimetic growth, i.e., to grow spatially oriented crystals from super saturated aqueous salts [39, 40].

The greatest advantage to work with Langmuir monolayers is that their structures and properties can be tuned easily by changing different physical and chemical properties such as surface pressure [29], temperature [27, 29], pH [32] and dissolved metal ion concentration [41].

It has also been observed that over and above pure organic monolayers, Langmuir monolayers, in presence of inorganic materials, makes the system rich both in structural and physical aspects. As an example, organic monolayers of long chain fatty acids in presence of divalent metal ions in the aqueous subphase show two-dimensional structural phases of the organic and inorganic parts at the air–water interface. Superlattice peaks in the diffraction pattern of fatty acid monolayer are observed in presence of cadmium ions in the water subphase [42, 43], as also in presence of magnesium, manganese and lead ions in the subphase [44, 45]. These pertain to two-dimensional ordered structures of the metal ions. On the other hand, the presence of certain metals (like cobalt, copper, nickel and barium) in the water induces a high pressure structure of the monolayer at low pressure without forming a lattice bearing the metal ions [45].

For certain experimental conditions new types of chemical bonds are also formed between hydrophilic part and metal ions [46]. Presence of metal ions in the subphase also makes the monolayer much more ordered [47], changes the viscoelastic properties [48] of the monolayer and helps the monolayer to transfer easily on to some solid substrate [49], i.e., in film deposition. In this book we have focused on some of these latter aspects of Langmuir monolayers and have demonstrated the effect of different metal ions in tuning the headgroup structure of fatty acid monolayers on water.

1.3 Langmuir–Blodgett (LB) Multilayers

1.3.1 Film Deposition

Deposition of multilayers of long chain carboxylic acid molecules onto a solid substrate, carried out by Katherine Blodgett [37], was the first instance of self-assembled growth of soft materials, and started the field of Langmuir–Blodgett film deposition. Conventionally, LB deposition is visualized as a process in which a solid substrate gets deposited with a monomolecular layer during each up/down stroke through a Langmuir monolayer of the amphiphilic molecules to be transferred, the organic multilayers formed on solid substrate having thicknesses equal to multiples of the Langmuir monolayer thickness. Out of the different processes by which monolayers and multilayers are formed on solid substrate from monolayers on water surface, films deposited by the *vertical* dipping of substrate are called *Langmuir Blodgett (LB) films* [5].

These films can be used to explore physics and chemistry in confined geometries and in designing model biological membranes [1, 3, 27, 47, 50, 51]. The molecular level control over the layer by layer growth in the film deposition makes this technique advantageous for preparation of an ideal system to study the two-dimensional to three-dimensional melting transition [52], two-dimensional magnetic properties [53, 54] and so on. Other than these basic interests, the LB films can also be used as templates for formation of the nanostructured materials [55]. CdS, PbS, ZnS quantum dots [56] or CdS nanosheets [57] are produced using the LB films. However, the mechanisms of film deposition or transfer from the water surface to the substrate surface and again to the deposited monolayer surface, are basically unexplained. These are of fundamental interest either as instances of wetting by complex liquids or as epitaxial growth through weak, short range forces.

1.3.2   Multilayer Structure

In LB deposition, a layer is deposited during each pass, such that molecules deposited on the upward pass have their polar head-groups oriented towards a hydrophillic substrate while molecules deposited on the downward pass are oriented with the hydrophobic tails towards the substrate. The order is reversed for hydropho-bic substrates. Irrespective of the type of substrate, however, the result is that, for amphiphilic molecules having one head and one tail, headgroups adhere to headgroups of the previous layer and tails stick to tails of preceding layer, to build up the multilayer.

There are several features common to multilayers created by this deposition technique. Firstly, the basic unit of an ideal LB multilayer is a *bilayer*, i.e., two monolayers adhered together by their heads with the tails pointing outward. Secondly, the first layer can be deposited only on the upward pass on hydrophilic substrates and on the downward pass on hydrophobic substrates. Thirdly, in air, hydrophobic tails are at the topmost layer while

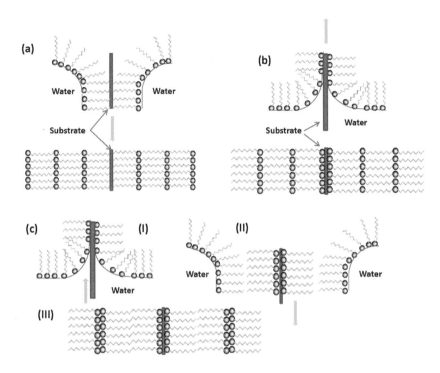

Figure 1.4: *(a) X-type, (b) Z-type, and (c) Y-type LB deposition.*

in water hydrophilic headgroups will be at the outermost layer. These three can be regarded as the basics of the LB film structure [47].

The energetics of amphiphilic interactions decide the engineering of structural modifications. For example, although monolayer transfer generally occur during both upward and downward pass (Y-type deposition), under certain less common conditions depending upon subphase pH, dissolved metal ions and hydrocarbon tail length, deposition can occur only during the downward pass (X-type deposition) or the upward pass (Z-type deposition). These three types of depositions are shown in Fig. 1.4. It is expected that for multilayer films formed using X-type or Z-type deposition the basic structural unit would be a monolayer instead of a bilayer. However, multilayers formed during X-type deposition have the usual bilayer structure as reported from x-ray diffraction studies

[58]. This implies that the molecules rearrange during or after transfer to attain a more energetically favorable configuration.

However, for fatty acid salts of multivalent metals, i.e., for amphiphilic molecules having more than one tail per headgroup, the situation is quite different and it is found that there are two different molecular configurations that forms the bilayer unit. One is the asymmetric configuration where all tails remain in same side of the metal-bearing headgroup and this forms the starting layer in LB deposition on a hydrophilic substrate. In the other, the attachment of two molecules is in symmetric configuration where tails remain in both side of the metal-bearing headgroups. Thus in symmetric configuration the attachment is due to tail-tail attachment. It has been observed by x-ray scattering experiments [59,60] that in all the layers, apart from the first layer on hydrophilic substrates, the molecular configuration is symmetric, i.e., tails are on both side of the headgroups as observed in crystals of these materials [61] whereas on hydrophobic substrates the entire multilayer is formed of molecules in this symmetric configuration. Using x-ray and neutron scattering, it has also been observed that for deposited fatty acid salts, even if the initial deposition is head to head, i.e., molecules are in asymmetric configuration, they rearrange themselves into the symmetric configuration [62–65]. We thus call the symmetric configuration of molecules the "symmetric monomolecular layer" or *"symmetric monolayer" (SML)* and the asymmetric configuration the *"asymmetric monomolecular layer" (AML)* [62]. These results are used frequently for building molecular models of LB films to analyse the deposited film structures, as we have done.

LB films are much more ordered in the out-of-plane direction, i.e., near-perfect one dimensional periodicity is maintained along the surface normal throughout the film thickness, while the in-plane domains of crystalline order extend to a maximum of micrometers, hence they are "powders" in plane [54,66,67], as found from grazing incidence x-ray diffraction. Within the crystalline domains of these LB films, the molecular packing of the constituent amphiphilic molecules depends upon the cations, alkane chain length and nature of substrate [47,68]. It is found that most

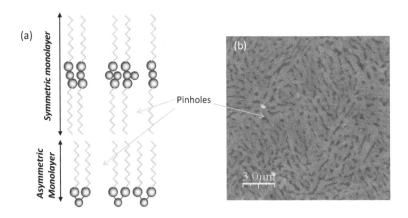

Figure 1.5: *(a) Schematic of LB film growth showing formation of "pinhole" defects and (b) "Pinholes" in cadmium stearate LB multilayers (AFM topographic image).*

LB films such as those of fatty acid salts always have some inherent defects. These defects are known as "pinhole defects" [47, 69] and they are some of the major causes that make the film less ordered in the two-dimensional plane of the headgroups. The interface morphology in LB films is the result between the competition of surface roughening from pinholes and the smoothening due to the surface tension of the layers [70]. Efforts have been devoted for obtaining better LB films free of pinhole defects. The most successful results have been obtained by Takamoto et al. [71] We have discussed them in details in this book and have shown that defect-free LB films can be produced by manipulating the headgroup structure via subphase metal ions.

1.4 Growth Mechanisms

Growth mechanisms associated with soft systems are less understood compared to metal and dielectrics, due to prevalence of polymorphism in the former. This has set a hurdle in reproducible fabrication of such soft materials as Langmuir–Blodgett films, which have been widely studied not only because of their applications in various areas [1] but also for the fundamentally new aspects of non-equilibrium processes involved. Most of the potential applications of LB films are based on premises of perfect molecular layering and orientation. Study of their growth mechanisms is thus of profound importance. Surface morphology, in plane correlations and growth mechanisms have been investigated by x-ray scattering and atomic force microscopy. Both self-affine and liquid-like correlations have been observed in plane [70,72,73]. One of the attempts is explaining the growth mechanism involved using linear stochastic theory where one-dimensional deposition followed by a two-dimensional desorption was the proposed mechanism [73]. However, a general unified approach is still lacking.

Based on the work of several groups, one can say that there are two approaches to soft materials growth: van der Waals local epitaxy and wetting by complex liquids. The first is due to Viswanathan et al. [74], in which growth of certain LB films is shown to proceed by a type of epitaxy called "strained-layer van der Waals epitaxy". It represents a compromise between strained-layer epitaxy and van der Waals epitaxy [75–77]. Viswanathan et al. predicted that the highest order in LB films results from optimization of a van der Waals type of epitaxy and that ordering in LB films, in general, may be achieved by designing molecules to maximize intralayer forces while minimizing layer-to-layer interactions [78,79] and simultaneously coupling the films to a substrate with similar lattice symmetry. However, what has not been carried out before is a systematic, layer-by-layer comparison between growth modes in LB multilayers, bearing metals, on one hand and the different growth modes encountered in general epitaxy on the other. In this book, we shall be demonstrating to what extent we have achieved this. In the next section, we have briefly discussed about crystal growth processes in general and also in the context of LB films. The major difference in the two cases is that whereas epitaxy in general refers to atom by atom growth, in LB films we refer to stacking of molecular layers, and we have to remember that intramolecular forces are orders of magnitude smaller than in case of inorganic materials.

The second approach is to consider LB deposition in the light of wetting and dewetting of fluids. Here, the growth of LB multilayers is considered to be wetting or dewetting of the hydrocarbon overlayer on the substrate [80]. This is decided by the forces of adhesion and cohesion and also by the interactions occurring at the metal–organic interface, the latter being decided by the mechanisms of complex formation. This is discussed later in this chapter.

Nevertheless, interconnection between molecular structure and growth of these films is not well understood. More specifically, the question as to why such systems prefer specific coordinations has not been answered. Our research has been an attempt to answer this question by studying a number of metal stearate LB films, deposited under identical conditions to understand headgroup structure of these films, and to find dependence of film morphology on headgroup structure. A major part in LB film research has been intent on controlling experimental parameters to obtain perfectly periodic LB films [47] but little effort has been made to manipulate headgroups to increase metal content and/or change the metal–headgroup interaction. Our work, in contrast, was focused on manipulating metal ion-headgroup interactions to change film morphology, and in the process finding out factors that are crucial, at molecular level, in dictating system morphology at microscopic scale and hence their growth.

1.5 Crystal Growth Processes

In this section, we define terminology associated with growth mechanism and briefly describe crystal growth processes in general.

1.5.1 Epitaxy

Epitaxy is crystal growth at a solid-liquid (or a solid-air) interface which acts as a matrix (template) for the bulk phase formation of material deposited further. Monolayer transfer is a special case of epitaxy since a two-dimensional lattice is preformed at the liquid-air interface. Peterson presented a detailed investigation of the conditions for epitaxy in transferred LB monolayers [2]. According to their criteria, the epitaxial growth of multilayers is characterized by the constancy of crystallographic packing, crystallite size, and orientation of the unit cell.

Homoepitaxy and Heteroepitaxy

Epitaxy thus refers to the method of depositing a monatomic layer on a crystalline substrate following the crystal structure of the latter. Homoepitaxy is a kind of epitaxy performed with only one material. In homoepitaxy, a crystalline film is grown on a substrate or film of the same material. This technology is used to grow a film which is purer than the substrate, as well as to fabricate layers having different doping levels. Heteroepitaxy, on the other hand, is a kind of epitaxy performed with materials that are different from each other. In heteroepitaxy, a crystalline film grows on a crystalline substrate or film of a different material. This technology is often used to grow crystalline films of materials for which single crystals cannot otherwise be obtained and to fabricate integrated crystalline layers of different materials.

Heteroepitaxial Growth Modes

There are three primary growth-modes (Fig. 1.6) by which thin films grow epitaxially at a crystal surface or interface. These are the Volmer–Weber, Frank Van der Merwe and Stranski–Krastanov growth modes [81, 82].

In Volmer–Weber (VW) or island-like growth-mode, small clusters nucleate directly on the substrate surface and build the base to islands of the condensed phase. This growth type appears when the molecules for deposition are more strongly bound to each other than to the substrate material.

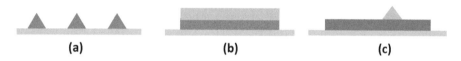

| (a) | (b) | (c) |

Figure 1.6: *Schematic of heteroepitaxial growth modes (a) Volmer–Weber, (b) Frank–Van der Merwe, and (c) Stranski–Krastanov.*

In contrast, Frank-Van der Merwe (FM) or the layer-by-layer growth mode happens in the opposite case, i.e., when the overlayer species (atoms or molecules) are more strongly bound to the substrate than to each other. In consequence the first atoms to condense form a complete monolayer (ML) on the surface, which is covered with further deposition with a slightly less bound second layer. For the case of continuing deposition of an increasing number of layers the binding energy shows a monotonic decrease towards the value of the bulk crystal of the deposit.

Stranski–Krastanov growth (SK growth, also Stransky–Krastanov or Stranski–Krastanow) is an intermediate case of these two primary modes in crystal-growth. Also known as "layer-plus-island growth," the SK mode follows a two step process: initially, complete films of adsorbates, up to several monolayers thick, grow in a layer-by-layer fashion on a crystal substrate. Beyond a critical layer thickness, which depends on strain and the chemical potential of the deposited film, growth continues through the nucleation and coalescence of adsorbate "islands."

Growth modes are of great technological interest as uniformity in thickness and perfection in crystallinity are of utmost importance in the fabrication of devices. Clearly an FM growth mode best serves the accomplishment of uniformity. The attainment of perfection in crystallinity is best served by growing epitaxially. It has been shown [83] that there exists a critical misfit, depending on relative bond strengths, below which an ML, satisfying the conditions for FM growth, is stable when homogeneously (misfit) strained into registry with the substrate, i.e., is pseudomorphic. As the thickness of the pseudomorphic film increases by continued deposition, the misfit strain energy increases and the pseudomorphic epilayer becomes unstable at a critical thickness $h' = h'_c$ (say) with respect to misfit strain relief by the introduction of misfit dislocations [84, 85]. When $h' > h'_c$, the misfit is accommodated by misfit dislocations and misfit strain jointly in the stable configuration, the misfit strain now being the homogeneous (average) part of the strain. It follows from the foregoing considerations that increasing epilayer multiplicity involves two different instabilities: (1) growth mode instability with a transition from FM to SK growth and (2) a misfit accommodation instability with a transition from pseudomorphic accommodation to disregistry by misfit strain release with the introduction of misfit dislocations. While it has been shown unequivocally in theory and experiment that the increase in strain energy with multiple layer formation is responsible for the latter instability, the often repeated suggestion [86, 87] that strain energy increase is the driving force for the growth mode instability and transition, is only a speculation and requires reliable quantification.

Bauer's Criteria

The phenomenon by which thin films commence to grow in different modes was first addressed by Bauer [88]. He formulated thermodynamic equilibrium criteria for the realization of growth modes, which have become generally accepted and will be generalized in the present approach.

While Bauer's criteria [88] can be derived in various ways, for example as a limiting case of Young's equation for contact angles, the present objectives are best served by a derivation employing equilibrium principles. Fig. 1.7 shows the double layer and ML sections of a growing epilayer; γ_0, γ_s, γ_i being respectively the specific surface free energies of the epilayer, the substrate and the epilayer-substrate interface. If transfer of material from the double layer to the ML decreases the free energy FM growth is favored, and otherwise VW growth. In terms of the free energy change ΔE_f, we have

$$\Delta E_f \equiv \Delta\gamma_0 = \gamma_0 + \gamma_i - \gamma_s \qquad (1.2)$$

$\Delta E_f \leq 0$ for FM growth, and $\Delta E_f > 0$ for VW growth. At this stage it is assumed that γ_0 and γ_i are independent of epilayer

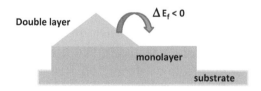

Figure 1.7: *Schematic of growing epilayer.*

thickness, i.e., is the same as for a bulk epilayer. The above equation not only assumes an equilibrated shape, i.e., quasi-equilibrium growth, but also somewhat drastic approximations, namely (a) that important proximity effects of the substrate may be neglected, (b) that MLs and double layers may be assigned macroscopic properties, e.g., of surface free energy and (c) that the equilibrium lattices of the ML and the double layer match that of the substrate, i.e., misfit strain and misfit dislocations are absent.

1.5.2   Growth Processes in LB Films

There exists no unified theory to describe the growth processes in metal–organic LB films. our research involves the study of LB films of transition-metal salts of fatty acids. Transition-metals, due to presence of highly directed orbitals of d-electrons [89], show a number of coordinations, particularly in their bonding with organic functional groups [90]. For each of these coordinations there may again be more than one conformation due to different energy states corresponding to alignment of d-orbitals relative to the coordinate bonds [91]. This gives rise to polymorphism in such organometallic compounds [92]. Metal carboxylates are simplest examples of these organometallics. But even in Langmuir–Blodgett (LB) multilayer growth [5] of divalent transition-metal carboxylates, different coordinations between metal and carboxylate moieties in the headgroup is observed [93], depending upon the particular transition-metal employed.

In order to understand the growth mechanism of these multilayers, it maybe worthwhile to recollect that the aforementioned "pinhole defects" observed in them are a common surface defect that occur, in particular, during manufacturing processes of one-dimensional periodic structures such as in epitaxial growth of device-grade crystals [94]. These defects that are common in Langmuir–Blodgett growth were, until recently, supposed to be inherent in this growth process [69]. In general, these defects are expected to occur due to intermolecular forces becoming stronger and longer-ranged than molecule-substrate forces. They are similar in nature to Volmer–Weber or island growth in heteroepitaxy in ultra high vacuum and for similar reasons, i.e., for them also the lattice mismatch is large, making interatomic forces stronger than atom-substrate interaction. However, relative importance of these long-range forces is not apparent in heteroepitaxial growth.

A major difference between organic i.e., molecular crystals and inorganic, i.e., metal, semiconductor or dielectric crystals lies in smallness of intermolecular forces in former. This smallness leads to incoherence in growth of these crystals and consequent lack of reproducibility. Thus controlling supramolecular interaction and, in particular, enhancing them by selecting specific molecular configurations should play a key role in growing functional organic crystals with a high degree of reproducibility. Studies on divalent transition elements interacting with organic materials relate to specific interconnections among coordination and bonding on one hand, and supramolecular interactions and morphology on the other, since such interconnections among structural hierarchy over many length scales provide the key to understand complex dynamics [95]. LB films of transition element salts of fatty acids are simplest of such complex systems, and hence many researchers, including us, have chosen these as model systems.

1.5.3 Growth in Specific Low-Dimensional Geometry

The inhomogeneous region between a bulk liquid and air, called the surface or interface, differs from bulk liquid in density and concentration of solvent and solute molecules. In addition, asymmetry in forces at interface can produce net orientation of molecules with respect to surface normal, unlike bulk, which is isotropic on average. As a consequence, properties of materials prepared at interface are different from those of the bulk. Comparative studies of reactions of soft organic materials at air/water interface versus those in bulk and consequent growth mechanisms associated in these cases are active areas of investigation. We have shown, by comparing a low-dimensional amphiphilc salt system with its bulk counterpart, that growth in these two cases are distinctly different and is governed by molecular geometry of these systems.

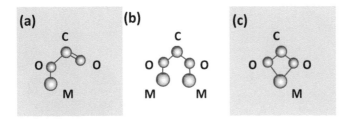

Figure 1.8: *Illustration of (a) unidentate, (b) bidentate bridged, and (c) bidentate chelate coordinations.*

1.5.4 Complex Formation at Air–Water Interface

In organic crystals, intermolecular forces are small compared to their inorganic counterpart. In general, when metal ions interact with any organic moiety, they do so by weak van der Waals forces to form metal–organic complexes. If the organic moiety is a fatty acid, the metal ion can interact with the polar headgroup (carboxylate ion) in three ways, giving rise to three types of molecular configurations or *coordinations*. These are (a) monodentate or unidentate; (b) bidentate bridge and (c) bidentate chelate coordinations. In unidentate coordination, one metal atom is attached to one of the oxygen atoms of the carboxylate group. Hence it is asymmetric and comparatively a less stable configuration. The bidentate coordination is symmetric in nature with the bridged coordination having two metal atoms attached to each of the oxygen atoms and the chelate having one metal atom attached to both the oxygen atoms. Due to its closed configuration, chelate is more stable compared to the bidentate bridged coordination. For each molecular coordination, there may exist more than one structural arrangements or molecular *conformations*. The vibrational energy of a molecule is decided by its coordination and conformation. We have studied and presented the possible coordinations and conformations formed due to interaction of different divalent metal ions with an amphiphilic fatty acid at air–water interface, and have discussed their role in deciding the growth of this type of metal–organics.

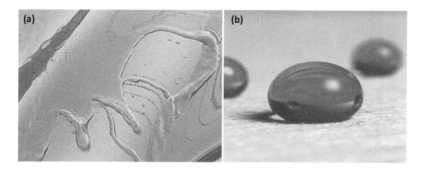

Figure 1.9: *Illustration of (a) wetting and (b) dewetting of liquid on a solid surface.*

1.6   Wetting by Complex Fluids

1.6.1   Wetting and Dewetting

Wetting is the ability of a liquid to maintain contact with a solid surface, resulting from intermolecular interactions when the two are brought together. The degree of wetting (*wettability*) is determined by a force balance between *adhesive* and *cohesive* forces. Adhesive forces between a liquid and solid cause a liquid drop to spread across the surface. Cohesive forces within the liquid cause the drop to ball up and avoid contact with the surface. Depending upon this force balance, the interface can be classified as wetting or dewetting.

1.6.2   Contact Angle

The contact angle (θ_{con}), as seen in Fig. 1.10, is the angle at which the liquid-vapor interface meets the solid-liquid interface. The contact angle is determined by the resultant between adhesive and cohesive forces. As the tendency of a drop to spread out over a flat, solid surface increases, the contact angle decreases. Thus, the contact angle provides an inverse measure of wettability [96]. A contact angle less than 90° (low contact angle) usually indicates that wetting of the surface is very favorable, and the fluid will spread over a large area of the surface. Contact angles greater

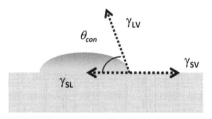

Figure 1.10: *Schematic of a liquid wetting a solid with a contact angle θ_{con}.*

than 90° (high contact angle) generally means that wetting of the surface is unfavorable so the fluid will minimize contact with the surface and form a compact liquid droplet. For water, a wettable surface may also be termed hydrophilic and a non-wettable surface hydrophobic. Superhydrophobic surfaces have contact angles greater than 150°, showing almost no contact between the liquid drop and the surface. Contact angles and their corresponding solid/liquid and liquid/liquid interactions have been reported by Eustathopoulus [97]. For non-aqueous liquids, the term lyophilic is used for low contact angle conditions and lyophobic is used when higher contact angles result. Similarly, the terms omniphobic and omniphilic apply to both polar and apolar liquids.

1.6.3   Young's Equation

Young's equation relates the surface tensions γ between the three phases: solid (S), liquid (L) and vapor (V). It is given by

$$\gamma_{SV} = \gamma_{SL} + \gamma_{LV} cos\theta_{con} \tag{1.3}$$

Subsequently it predicts the contact angle of a liquid droplet on a solid surface from knowledge of the three surface energies involved. This equation also applies if the "vapor" phase is another liquid, immiscible with the droplet of the first "liquid" phase.

1.6.4   High Energy and Low Energy Surfaces

There are two main types of solid surfaces with which liquids can interact. Traditionally, solid surfaces have been divided into high

energy solids and low energy types. The relative energy of a solid
has to do with the bulk nature of the solid itself. Solids such as
metals, glasses, and ceramics are known as "hard solids" because
the chemical bonds that hold them together (e.g., covalent, ionic,
or metallic) are very strong. Thus, it takes a large input of energy
to break these bonds to create the surface of these solids so the
surfaces are termed "high energy." Most molecular liquids achieve
complete wetting with high-energy surfaces to minimize surface
free energy. The other type of solids are molecular crystals (e.g.,
fluorocarbons, hydrocarbons, etc.) where the molecules are held
together essentially by much weaker physical forces (e.g., van der
Waals and hydrogen bonds). Since these solids are held together
by weak forces it would take a very low input of energy to create the surface, and thus, they are termed "low energy" surfaces.
Depending on the type of liquid chosen, low-energy surfaces can
permit complete or partial wetting as well as dewetting [98, 99].

1.6.5   Wetting of Low Energy Surfaces

Low-energy surfaces primarily interact with liquids through dispersion (van der Waals) forces. William Zisman had several key
findings in this regard [100]. Zisman observed that (1) $\cos \theta_{con}$ increases linearly as the surface tension γ_{LV} of the liquid decreased.
Thus, he was able to establish a rectilinear relation between $\cos$
θ_{con} and γ_{LV} for various organic liquids. (2) A surface is more
wettable when γ_{LV} is low and when θ_{con} is low. He termed the
surface tension when $\cos \theta_{con} = 1$, as the critical surface tension
(γ_c) of that surface. This critical surface tension is an important
parameter because it is a characteristic of only the solid. Know-
ing the critical surface tension of a solid, it is possible to predict
the wettability of the surface [96]. (3) The wettability of a surface is determined by the outermost chemical groups of the solid.
(4) Differences in wettability between surfaces that are similar in
structure are due to differences in packing of the atoms. For instance, if a surface has branched chains, it will have poorer packing
than a surface with straight chains.

1.6.6 Young–Dupré Equation and Spreading Coefficient

A useful parameter for quantification of wetting is the spreading co-efficient S,

$$S = \gamma_{SV} - (\gamma_{SL} + \gamma_{LV}) \qquad (1.4)$$

When $S > 0$, the liquid wets the surface completely (complete wetting). When $S < 0$, there is partial wetting. Combining the spreading parameter definition with the Young relation, we obtain the Young–Dupré equation as

$$S = \gamma_{LV}(cos\theta_{con} - 1) \qquad (1.5)$$

which only has physical solutions for θ_{con} when $S < 0$.

1.6.7 Effect of Surfactants on Wetting

Many technological processes require control of liquid spreading over solid surfaces. When a drop is placed on a surface, it can completely wet, partially wet, or not wet the surface. By reducing the surface tension with surfactants, a non-wetting material can be made to become partially or completely wetting. The excess free energy (σ_e) of a drop on a solid surface is given by [101],

$$\sigma_e = \gamma_{LV} + \pi R_d^2 \gamma_{SL} - \gamma_{SV} \qquad (1.6)$$

where R_d is the radius of droplet base, and γ is surface tension.

The excess free energy is thus minimized when γ_{LV} decreases, γ_{SL} decreases, or γ_{SV} increases. Surfactants are absorbed onto the liquid-vapor, solid-liquid, and solid-vapor interfaces, which modify the wetting behavior of hydrophobic materials to reduce the free energy. This is possible because surfactants or amphiphiles, unlike other liquids, comprise both hydrophillic and hydrophobic parts that suitably interact with the substrate to reduce their surface energy. Hence, their wetting and dewetting behavior is relatively unknown, as de Gennes has pointed out [102], and may lead to surprises such as the spontaneous formation of nanocrystals, as we have found out.

1.7   Outline of Book

In this book we have tried to correlate the structural and morphological aspects of Langmuir monolayers and Langmuir–Blodgett films in presence of different divalent transition metal ions. On one hand, we have tried to explore the molecular structure of these organic films and determine factors, especially changes in bonding, which can possibly control them. On the other, we have tried to evaluate the growth mechanism of these films by studying their morphological evolution at the microscopic scale. We have correlated the growth to the molecular structure, and examined the change in the former by tuning the latter. In the process, we have found out unique wetting and dewetting properties of these organic films and their ability to behave as tunable low energy templates for organic film growth.

We have already made a brief review on the subject in this chapter. In the next chapter, we have outlined the sample preparation and and then in the third chapter, the probing techniques. We have presented the theoretical background for extracting information from the different probing techniques in Chapter 4. Chapters 5 and 6 discuss the association of morphology with structure in Langmuir Monolayers and Langmuir–Blodgett multilayers, with cadmium stearate and cobalt stearate films as examples and an emphasis on the crucial role played by air–water interface during transfer. Chapter 7 includes the classification of four different metal-fatty acid films on the basis of their morphological evolution on one hand and determination of factors that control the growth mechanism in such soft systems on the other. In Chapter 8, we have presented the behavior of cadmium stearate and cobalt stearate trilayer films as two-dimensional "solids" and "liquids." We have looked at their dewetting properties and showed their use as templates for nanocrystal self-assembly of zinc stearate. In the last chapter, we have drawn some conclusions from our research and possible outcomes.

Chapter 2

Preparing Monolayers and Multilayers

In this book we present methods for preparing monolayer and multilayer films that depend on formation of the films on water surface and their transfer onto solid substrates.

2.1 Preparing the Monolayer on Water

2.1.1 Pristine Monolayers

In Chapter 1, we have related the general outlines of how the Langmuir monolayer is formed by spreading it on water. Some materials spread spontaneously on water but most need a spreading solvent. It is desirable that the solvent should be capable of dispersing the molecules of the film-forming material at the air–water interface and then evaporate completely so that the film is not contaminated. Hence, the spreading solvent chosen is volatile and insoluble in water. Typical examples are chloroform, methanol, benzene or a suitable mixture of these solvents.

A volatile solvent which is intended to evaporate must do so within a reasonably short time; extremely volatile solvents, however, present difficulties because evaporation prevents accurate determination of solution concentrations. Solvents which have boiling point in the range 40°C-80°C are generally most suitable for experiments under ordinary conditions. High water solubility should definitely be avoided since some of the film-forming material may be carried into the aqueous subphase and may be precipitated rather than appearing on the surface.

If a drop of liquid (1) is placed on a solid or another liquid (2), the extent to which it spreads is given by Young's equation (Eqn. 1.3). For complete wetting there is no non-zero contact angle, i.e., the liquid spreads completely on the solid or liquid. The *spreading coefficient* S_{12}, as defined in Chapter 1 (Eqn. 1.4), is a measure of the *wettability* of liquid 1 on liquid/solid 2. If the value of S_{12} is positive, spreading will occur, while if it's negative, liquid 1 will dewet and hence rest as a lens on liquid 2.

In general, liquid amphiphiles or a concentrated solution of amphiphiles in any solvent has a negative value of S_{12}, where 1 stands for the solution and 2 stands for water. For a solution of amphiphiles to spread uniformly over water and to form a monolayer, a dilute solution of the amphiphiles in a solvent that has a large positive S_{12} value should be formed. The solvent must, of course, have sufficient dissolving power for the amphiphile, be chemically inert with respect to the film material and should be easily purified to eliminate contaminants.

The spreading solution is added drop-wise on different parts of the water surface from some micro-syringe containing a desired amount of the solution and then some time is allowed for the complete evaporation of the solvent.

The amount of solvent spread decides the initial average area available to each molecule in the monolayer. If M_{am} is the molecular weight of the amphiphile, ρ_A is the concentration of the amphiphilic solution in g ml^{-1}, V_S is the volume of the solution spread in ml and the total surface area covered by the monolayer is A_{ML} cm^2, then the area per molecule A is $A_{ML}M_{am}/\rho_A V_S N_A$ cm^2 or $(A_{ML}M_{am}/\rho_A V_S N_A) \times 10^{16}$ Å^2, where N_A is Avogadro's number. Hence the concentration and volume of solution spread is chosen so that the area per molecule is large enough for the monolayer to be in the gas phase.

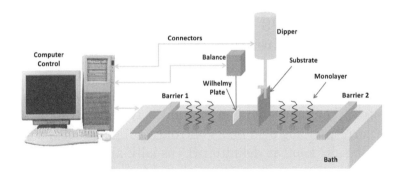

Figure 2.1: *Schematic of a Langmuir trough.*

2.1.2   The Langmuir Trough

The instrument used for preparation and measurement of Langmuir monolayers is called a trough. The basic schematic of a single bath Langmuir trough is shown in Fig. 2.1. The essential elements of that trough are (a) the bath, usually made of a hydrophobic material like teflon, (b) the mobile barriers for controlling the surface area, (c) a balance for measuring the surface pressure, and (d) a dipper for dipping the substrate in the monolayer for Langmuir–Blodgett film deposition. The bath is filled with the subphase , usually ultrapure water of high resistivity. The mobile barriers are moved to the sides of the trough which corresponds to maximum decompression position (*open trough configuration*). A very dilute solution of the amphiphile of interest is introduced at the surface with the help of a micro syringe, and the volatile solvent is allowed to completely evaporate. The monomolecular layer of amphiphiles is then compressed by bringing together the two mobile barriers at a slow constant rate, whence the monolayer passes through different phases as shown in Fig. 1.3. The isotherm is recorded by measuring the value of surface pressure π_s for each barrier position as a function of the area per molecule.

2.1.3   Preparing Monolayers of Salts: Example of 2D Chemical Reaction

There are two different ways in which monolayers of salts of long chain fatty acids, the prototypical amphiphilic molecules, may be deposited. The most common method is to spread the monolayer of the corresponding acid on a controlled subphase, of the suitable inorganic salt, kept at definite pH (usually alkaline) [26]. In this method, the reaction occurs in one step, i.e., the fatty acids, when floating in a Langmuir monolayer on water surface, can react directly with ions of the inorganic salt, dissolved in water in less than μM concentrations to yield a nearly purified form of the salt. In the other method, the salt is actually prepared beforehand i.e., preformed and spread on water at normal pH (around 5.65). Bulk preparation of these type of metal carboxylate salts is a two-step process. The required ions do not interact directly with the acid, the process involving preparation of salt of the fatty acid with an alkali metal and then by an exchange reaction, the divalent metal forms the salt [65, 109]. Let us have a clear idea of the latter process with a specific example of forming cobalt stearate from stearic acid.

In the first step of preparation of cobalt stearate, sodium stearate has to be prepared by adding sodium hydroxide to pure (Milli-Q) water containing stearic acid in appropriate amounts at about 80°C - 90°C. Addition of sodium hydroxide was continued until the medium was slightly alkaline (pH 7.0-7.5). Sodium stearate was completely soluble in hot water. In the second step, we added equivalent amount of cobalt chloride solution to the freshly prepared sodium stearate solution in hot condition, whence cobalt stearate was formed as an insoluble blue flocculate on water, and was filtered out. In the third step, we repeatedly washed the cobalt stearate with hot pure water to filter out any unreacted sodium stearate and other water soluble impurities, followed by a repeated wash with benzene to remove any unreacted stearic acid and other organic impurities.

2.2   Preparing the Film on Substrates

Films of amphiphilic molecules, built up by passing a solid substrate through a Langmuir monolayer, are known as Langmuir–Blodgett films. These can be single layers or multilayers depending on the number of passes made through the water surface. The substrates used to deposit these films are specially treated for good transfer, e.g., they are made hydrophilic or hydrophobic depending on the type of deposition. LB deposition is carried out in the Langmuir trough. The substrate is attached to a dipper, and passed through the air/water interface on which the Langmuir monolayer of the material to be deposited is formed and maintained under required experimental conditions. There are two types of monolayer transfer: the vertical type and horizontal type, depending on the substrate orientation with respect to the interface. Both techniques are discussed below. However, for multilayer deposition, one has to use the vertical technique.

2.2.1   The Langmuir–Blodgett Technique of Vertical Deposition

The process of building an LB multilayer consists of passing the film alternately upwards and downwards through a Langmuir monolayer at the air/water interface. These upstrokes and downstrokes comprise the technique of vertical deposition. As discussed in Chapter 1, a layer is deposited during each pass; viz. the upward pass ("upstroke") and the downstroke. For monolayers of amphiphiles having one headgroup and one tail the result is that the headgroups adhere to the headgroups of the previous layer during the upstroke and the tails stick to the tails of the previous layer during the downstroke, in accord with what is expected from amphiphilic molecules. The process is illustrated in Fig. 2.2. The energetics of the amphiphilic interactions severely limit the ability to engineer structures one might want. For example, although monolayer transfer generally occurs during both upstroke and downstroke (Y-type deposition), under certain less common conditions, deposition can occur only during the downstroke (X-

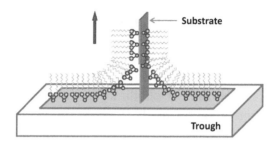

Figure 2.2: *Schematic of vertical LB deposition.*

type) or the upstroke (Z-type) as explained in Chapter 1. One
might expect that by depositing a multilayer using X-type or Z-
type deposition, one could manufacture films with the basic structural unit being a monolayer; that is, the molecules in every layer
would have the same orientation. However, using x-ray diffraction, it is shown that multilayers built by X-type deposition, have
the usual symmetric monolayer structure [58] implying that the
molecules rearrange during or after transfer to attain a more energetically favorable configuration. As another example, one might
try to prepare a film with polar headgroups sticking outward in
air by removing the Langmuir monolayer from the water surface
before the final upstroke. In this situation, however, the outer
layer is simply "ripped off" from the film as it passes through the
surface [37, 47], leaving the previous layer with the tails facing
outward. During deposition of an individual LB layer, the surface
area of the Langmuir monolayer on the water surface, held at constant surface pressure, decreases due to the loss of molecules to
the substrate. A simple diagnostic parameter of the deposition,
the transfer ratio (*TR*), is defined as

$$TR = \frac{Decrease\ in\ surface\ area\ of\ Langmuir\ monolayer}{Surface\ area\ of\ substrate}$$

(2.1)

According to conventional wisdom, a *TR* of unity is indicative of
good deposition and, in fact, values near unity are often observed
under certain conditions. The acceptance of the value of unity
as ideal, however, reflects a strongly ingrained implicit assump-

tion that the deposition process consists of simply transferring the molecules from the water surface to the substrate and ending in the creation of a sort of replica. However, this assumption is too simplistic and is often incorrect [47]. LB transfer is a complicated process in which the amphiphiles generally attempt to reach a new equilibrium as they experience a change in interactions from those with water surface to those with the solid substrate. If the molecular packing density changes during transfer, then $TR = 1$ will not be the indicator of a defect-free film. In many cases, however, the molecules on the water surface are compressed into a nearly close-packed state before transfer and, although the details of the molecular packing may change, the molecules remain densely packed after transfer. In such cases, the difference in packing density before and after will be within a few percent of the ideal value, consistent with the typical uncertainty of TR measurement.

2.2.2 Modified Inverted Langmuir–Schaefer Method for Horizontal Deposition

The monolayer transfer process in the conventional LB deposition does not work either for viscous and rigid films or for films at very low surface pressures. Alternative deposition processes have been considered to overcome this problem. In particular three different horizontal deposition techniques have yielded good results. In the first deposition scheme employed by I. Langmuir and V. J. Schaefer (LS method) [110], the substrate is always hydrophobic in nature and the film is "lifted by the tails" from the top. The substrate is aligned parallel to the monolayer surface and the monolayer is transferred when the substrate touches the monolayer from the top. However, frequent overturning [111] of molecules has also been observed during this deposition process. An important limitation of both LB and LS deposition technique is that the local molecular organization, the distribution or size of domains and the fraction of various phases can change during the transfer.

The second technique is inverse Langmuir Schaefer (ILS) method [112] where substrate must be hydrophilic and the film is lifted onto the substrate by holding it underneath the film on

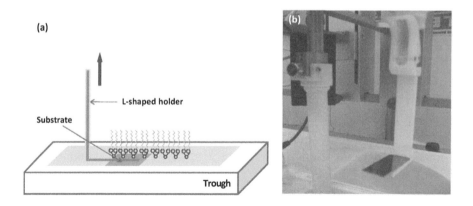

Figure 2.3: *(a) Schematic of MILS method for horizontal deposition where an L-shaped teflon substrate holder is attached to the dipper of the Langmuir trough, and (b) close view of holder.*

water and draining out the water at a prescribed rate. This technique makes possible the transfer of rigid films. It also requires a special kind of trough with an arrangement for highly controlled removal of water. The third technique was used by Kato [113]. In this technique hydrophilic solid substrate was kept almost horizontal just beneath the water surface before the monolayer was spread, and after achieving the desired condition of the monolayer, the substrate was raised up slowly keeping it horizontal. Following previous work done from our laboratory [63], we have used this modified ILS method of Kato [114], where instead of water drainage, the substrate holder moves slowly from water to air. This modified inverted Langmuir Schaefer (MILS) method is simple and can be used very easily with any commercial LB trough. The schematic of the MILS method is shown in Fig. 2.3. At the time of film deposition in the MILS method, the water surface was properly cleaned, and a home-made L-shaped teflon substrate holder was attached to the LB trough dipper and immersed into the water such that the substrate is parallel to and ~10 mm below the air–water interface. Then the film was spread on water and desired surface pressure was maintained by compression. The substrate holder can be taken out from water to air with desired speed using the dipper mechanism. All the substrates which were used for film deposition in MILS method are hydrophilic in nature, and in fact, we have always used hydrophilic deposition.

2.2.3   The Alternating Trough

While in a conventional single bath trough there is only one dipping arm for dipping the substrate, in the alternate trough (Fig. 2.4) there are two dipper arms. The advantage of this trough as compared to the ordinary trough with a single bath is that it can be used to deposit two different types of materials in alternate cycles to form an ABAB... type film. This system has two baths with independent sets of two barriers which are hydrophilic to ensure that the film material does not go under the barriers during compression. The maximum and minimum speeds of the barriers, controlled by a micro-step driven stepping motor, are typically 0.5 and 85 mm/min, respectively. In addition, each bath contains a Wilhelmy plate made of sandblasted platinum. There is an elevator to position both these plates inside the subphase simultaneously, the usual practice being to submerge one third of the plates inside the subphase. The upper arm is used to mount the substrate initially and for bringing it upto the subphase while

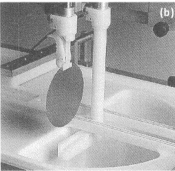

Figure 2.4: *KSV 5000 Alternating Trough : (a) front view and (b) close view of the dipper.*

the lower arm is used to transport the substrate inside the sub-phase and between the two baths. The upper arm never goes below the subphase while the lower arm always remains within the subphase. The dipping speed of the substrate can be varied from 0.5 to 85 mm/min. The dipper arms are controlled by in-dependent stepper motors. If pure water is used as the subphase it should have resistivity of 18 MΩ cm. The temperature of the subphase can be regulated by circulating a suitable liquid from an external temperature controlled water bath inside the circulator fitted beneath the trough. To deposit an LB film the precursor monolayer is compressed to a pressure such that the monolayer is in the solid region of the $\pi_s - A$ isotherm (Fig. 1.3). During deposition the monolayer pressure is held constant by compensat-ing for the material transferred onto the substrate in terms of the reduction of the area occupied by the monolayer.

Chapter 3

Probing Tools

Probing Langmuir monolayers and multilayers (formed by LB, MILS or other methods) now uses a large arsenal of experimental techniques starting from simple measurements of surface tension or pressure to surface sensitive spectroscopy, scattering studies and microscopy. In this chapter we will give a brief overview of the working principle of some of these techniques used by us.

3.1 Isotherm Measurements

The most common experiment performed to check the formation and stability of Langmuir monolayers and study its mechanical properties is the determination of surface pressure (π_s) versus specific molecular area (A) isotherms [25, 26]. Upon decreasing the area of the monolayer, the surface tension of the monolayer-covered water surface (γ) decreases and hence π_s increases as the surface tension of the pure water (γ_w) is constant (see Eqn. 1.1). This surface pressure, which is roughly equal to S_{12} for small γ_{21} thus changes with change in area per molecule of the amphiphilic monolayer. There are two different experimental techniques by which π_s-A isotherm can be obtained, viz. Wilhelmy balance and Langmuir balance.

The Langmuir balance method is basically a direct differential measurement. A clean portion of the liquid surface is separated from the film-covered area by a partition, and it is the force acting on this partition which is measured. The partition consists of a movable float connected to a conventional balance with which the magnitude of the force is determined. The Wilhelmy balance, on the other hand, is more commonly used and the working principle of the Wilhelmy balance is given below.

3.1.1   Wilhelmy Balance

In the Wilhelmy method, an absolute measurement is made by determining the force due to surface tension on a plate or other object suspended so that it is partially immersed in the Langmuir monolayer, and this is compared with a similar absolute measurement on a reference surface, e.g., a clean water surface.

In the common form of the Wilhelmy [103] balance, a thin plate, usually of platinum, is suspended in the liquid surface as shown in Fig. 3.1. The forces on the plate then consist of the gravitational and surface tension effects, acting downward, partially offset by the buoyant effect due to the weight of liquid displaced. For a rectangular plate of dimensions $l \times w \times t$, of material density ρ_p, immersed to a depth h_p in a liquid of density ρ_l, we have for the net downward force

$$|F_{WB}| = \rho_p g_a lwt + 2\gamma_l(t+w)\cos\theta_{con} - \rho_l g_a twh_p \qquad (3.1)$$

where γ_l is the liquid surface tension, θ_{con} is the contact angle of the liquid on the solid plate, and g_a is the acceleration due to gravity. This equation neglects the second phase, which is ordinarily air with negligible density. It can be generalized readily to include two-liquid interfaces, in which case γ_l is the liquid-liquid interfacial tension. It should also be noted that a slightly more rigorous analysis [104] can be made on the basis of the weight of liquid raised in the meniscus on the plate as shown in Fig. 3.1. This avoids the difficulty in energy balance involved in evaluating the contact angle for complete wetting.

The usual procedure for the use of a Wilhelmy balance involves maintaining the plate completely wetted by the liquid ($\cos\theta_{con} = 1$) and measuring either the change in F_{WB} for a stationary plate

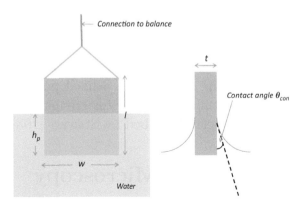

Figure 3.1: *Schematic of a Wilhelmy balance partially immersed in water.*

or the change in h_p for a constant applied force, when the surface tension is altered. In the former case, Eqn. 1.1 yields

$$\pi_s = -\Delta\gamma = -[\Delta|F_{WB}|/2(t+w)] \tag{3.2}$$

or if the force is constant,

$$\pi_s = -\Delta\gamma = -[\rho_l g_a t w/2(t+w)]\Delta h_p \tag{3.3}$$

In the latter case, if the plate is thin enough so that t is negligible compared to w,

$$\pi_s = -(1/2)\rho_l g_a t \Delta h_p \tag{3.4}$$

from which it is apparent that the sensitivity can be increased greatly by using a very thin plate.

Various methods have been used for measuring the forces involved with a Wilhelmy plate. Harkins and Anderson [105] first gave a detailed analysis of the application of the method, used simple beam balances. The plate is connected to one side of the balance beam, and the system balanced with the plate partly immersed in a clean water surface. Movements of the plate are then observed by noting deflections of the balance, e.g., with a spot of light reflected from a mirror on the beam.

In Dervichian's instrument [106], the reflected light was directed onto a moving piece of sensitized paper to give a direct record of film pressure. A more modern variation involves the use of an electrically actuated recording balance, which has the advantage of maintaining the plate in a fixed position in the interface [107]. A capacitance null detector circuit suitable for use with such devices has been described by Stewart [108].

3.2   Atomic Force Microscopy

Atomic force microscopy (AFM) is a very powerful tool for studying surface topography as well as surface properties. It is a member of the family of *scanning probe microscopy* (SPM), which gives three dimensional images of surfaces at the atomic scale [115–117]. SPM is based on scanning a probe or *tip* within a few nanometers above the surface of interest whilst monitoring some interaction between probe and surface. The interaction is spatially localized according to shape of probe and finite effective range of interaction.

When the probe moves laterally (horizontally) relative to the sample, any change in height of the surface causes the tip-surface interaction to change. By keeping the interaction strength at a constant preset value with a feedback circuit, it is possible to guide the tip at a finite tip-surface separation over the sample, exploiting this height-dependence of the tip-sample interaction. By raster scanning the tip, surface contours of constant interaction strength are obtained. Changing the type of interaction generally leads to changes of the surface contours obtained. Alternatively, the tip can be raster scanned at a fixed height over the sample surface and variations of the interaction strength, resulting from variations of tip-to-sample spacing, can be recorded.

The resolution varies somewhat from technique to technique, but some probe techniques reach a rather impressive atomic resolution. They owe this largely to the ability of *piezoelectric transducers* to execute motions with a precision and accuracy at the atomic level or better on electronic command. One could rightly call this family of techniques "piezoelectric techniques." The other common denominator is that the data are typically obtained as a two-dimensional grid of data points, visualized in false color as a computer image.

Many scanning probe microscopes can image several interactions simultaneously. The manner of using these interactions to obtain an image is generally called a mode. Some these modes are AFM or *scanning force microscopy* (SFM), where tip-sample interaction is predominantly the van der Waals force; *Scanning tunneling microscopy* (STM), where tunneling current between a metallic tip and a conducting substrate is probed; *Magnetic force microscopy* (MFM), which measures magnetic force interaction between tip and sample; *Lateral force microscopy*(LFM), that detects the torsion of the scanning tip; *Electrostatic force microscopy*(EFM) where long range electrostatic Coulomb forces are probed, and so on. In our book, we have described the use of AFM to study the morphology of films on substrates.

3.2.1 Basic Principle

In AFM the predominant tip-sample interaction is van der Waals force . For two electrically neutral and non-magnetic bodies held at a distance (r) of one to several tens of nanometers, the van der Waals force $F_{VW}(r)$ can be derived from Lennard–Jones potential $V_{LJ}(r)$ given by,

$$V_{LJ}(r) = 4\epsilon_d \left[\left(\frac{\sigma_0}{r} \right)^{12} - \left(\frac{\sigma_0}{r} \right)^{6} \right] \quad (3.5)$$

where ϵ_d is the depth of the potential well and σ_0 is the finite distance at which the inter-particle potential is zero.

The distance dependence of $F_{VW}(r)$ can be written as

$$F_{VW}(r) = 4\epsilon_d \left[\left(\frac{\sigma_0}{r} \right)^{13} - \left(\frac{\sigma_0}{r} \right)^{7} \right] \quad (3.6)$$

The interatomic force versus distance curve is shown in Fig. 3.2. As the atoms are gradually brought together the attractive force

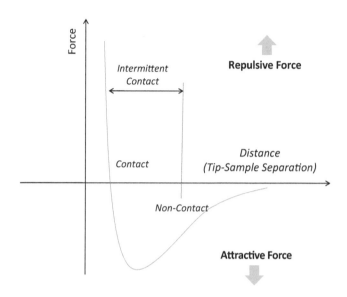

Figure 3.2: *Interatomic force versus distance curve.*

between them due to mutually induced dipole moments increases until they are so close together that their electron clouds begin to repel each other electrostatically. This electrostatic repulsion progressively weakens the attractive force as the interatomic separation continues to decrease and when the distance between the atoms reach a couple of angstroms, the total van der Waals force becomes repulsive.

In AFM, the local variation of the force acting between the tip and the sample is measured in order to generate the 3D images of the surface. In Fig. 3.3 the operating principle of AFM is shown with the help of a schematic diagram. AFM relies on the use of a sharp tip mounted on a cantilever-type spring which is brought into close proximity to the surface and as the tip moves over the surface the change in local interaction forces acting between them cause the cantilever to bend according to Hooke's law. The cantilever deflection is detected by "optical lever" principle and converted into an electrical signal to produce the image. In optical lever method, a laser beam reflected from the backside of the cantilever is made to be incident on a position sensitive photo detector

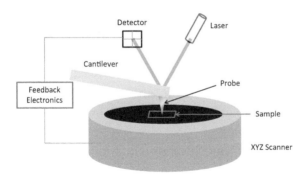

Figure 3.3: *Schematic diagram of AFM showing the basic principle of operation.*

(PSPD). As the cantilever deflects, the angle of the reflected beam changes and the spot falls on a different part of the photodetector. Generally the detector is made of four quadrants and the signals from the four quadrants are compared to calculate the position of the laser spot. The vertical deflection of the cantilever can be calculated by comparing the amount of signal from "top" and "bottom" half signals from the detector. The lateral twisting of the cantilever can also be calculated by comparing "left" and "right" half signals from the detector. This detection system measures the cantilever deflection with sub-angstrom sensitivity. The spring constant of the cantilever should be small enough to allow detection of small forces and its resonant frequency should be high to minimize sensitivity to mechanical vibration. The scanning of the tip or sample movement is performed by an extremely precise positioning device made from piezo-electric ceramics, most often in the form of a tube scanner. The scanner is capable of sub-angstrom resolution in x, y and z-directions. To control the relative position of the tip with respect to the sample accurately, active vibrational isolation of the microscope has to be ensured.

3.2.2   Modes of Operation

There are three main modes of operation of an AFM - contact mode, tapping mode and non-contact.

Contact Mode

In contact mode the tip is always in the repulsive regime of the intermolecular force curve as shown in Fig. 3.2. The forces range from nano to micro N in solids and even lower (0.1 nN or less) in liquids under ambient conditions. In this mode imaging can be done in two ways:

(i) **Constant Force Imaging** in which a feedback circuit adjusts the height of the tip during scanning so that the cantilever deflection and hence the corresponding force remains constant. The feedback system is connected to the z-piezodrive and the output signal (U_z) of the feedback loop adjusts the vertical z position of the sample or tip to achieve a constant cantilever deflection (constant force). U_z can be recorded as a function of the (x, y) coordinates which are determined by the corresponding voltages U_x and U_y applied to the x and y piezoelectric drives. The obtained signal $U_z (U_x, U_y)$ can finally be translated into the "topography" $z (x,y)$, provided that the sensitivities of the three orthogonal piezoelectric drives are known.

(ii) **Constant Height Imaging** in which tip is scanned at a fixed height above the sample and the variation in tip-sample interaction force reflects the atomic-scale topography of the sample. This mode is useful when the tip has to be scanned faster than the finite response time of the feedback loop. A significant drawback of this mode lies in the fact that vertical height information is not directly available. Although the constant force mode is generally preferred for most applications, constant height mode is often used for taking atomic-scale images of atomically flat surfaces, where the cantilever deflections and thus the variation in force are small. This is also essential for recording real-time images of changing surfaces, where high scan speed is essential. A significant drawback of contact mode is that the dragging motion of the probe tip, combined with adhesive forces between the tip and the surface, can cause substantial damage to both sample and probe and create artifacts in the image. Thus this mode is not an ideal one to perform high resolution topographic imaging of surfaces of soft materials which can easily be damaged or of overlayers those are loosely held to their substrates.

Non-Contact Mode

Non-contact AFM (NC-AFM) is a vibrating cantilever technique in which an AFM cantilever is vibrated above the surface of the sample at such a distance that we are no longer in the repulsive regime of the inter-molecular force curve as indicated in Fig. 3.2. Since in non-contact regime, there is no contact between the tip and the sample and the total force acting between them is very low, this mode is advantageous for studying soft or elastic samples. Cantilever used here must be stiffer than those used for contact AFM because soft cantilever can be pulled to contact with the sample surface. The small force values in non-contact regime and the greater stiffness of the cantilevers make the NC-AFM signal small and therefore difficult to measure. Thus a sensitive AC detection scheme is used for NC-AFM operation, where a stiff cantilever vibrates slightly above its resonant frequency (typically from 100 to 400 kHz) with an amplitude of a few tens to hundreds of angstroms. The resonant frequency of the cantilever is decreased by van der Waals force resulting in a change in oscillation amplitude. The system monitors the changes in resonance frequency or vibrational amplitude of the cantilever and these changes can be used as a measure of changes in the force gradient, which reflects changes in the tip-to-sample spacing or sample topography. The resonance frequency or vibrational amplitude of the cantilever is kept constant with the aid of a feedback system that moves the scanner up and down. By keeping the resonant frequency or amplitude constant, the system also keeps the average tip-to-sample distance constant. As with contact AFM, the voltage applied to the scanner U_z (U_x, U_y) is used to generate the topography. Although this mode does not suffer from the tip or sample degradation effects that are sometimes observed after taking several scans with contact AFM, contact AFM often provides better resolution than NC-AFM. This is because the attractive van der Waals forces are substantially weaker than the forces used by contact mode and the attractive forces extend only a small distance from the surface, where the adsorbed fluid layer may occupy a large fraction of their useful range. Hence, even when the sample-tip separation is successfully maintained, non-contact mode provides substantially lower resolution than contact mode.

Tapping Mode

Tapping mode imaging overcomes the limitations of the conventional scanning modes by alternately placing the tip in contact with the surface to provide high resolution and then lifting the tip off the surface to avoid dragging the tip across the surface. Tapping mode imaging is implemented in ambient conditions by oscillating the cantilever assembly at or near the cantilever's resonance frequency using a piezoelectric crystal. The piezo motion causes the cantilever to oscillate with a high amplitude (typically greater than 20nm) when the tip is not in contact with the surface. The oscillating tip is then moved toward the surface until it begins to lightly touch, or "tap" the surface. During scanning, the vertically oscillating tip alternately contacts the surface and lifts off, generally at a frequency of 50,000 to 500,000 cycles per second. As the oscillating cantilever begins to intermittently contact the surface, the cantilever oscillation is reduced due to energy loss caused by the tip contacting the surface. This mode can be used to obtain topographic as well as phase images.

(i) **Topographic Imaging:** The amplitude of oscillation changes when the tip scans over the bumps or depressions. A constant oscillation amplitude and thus a constant tip-sample interaction is maintained during scanning with the help of the feedback circuit by adjusting the tip-sample separation, giving us the topographic image of the surface. This mode inherently prevents the tip from sticking to the surface and causing damage during scanning. Unlike contact and non-contact modes, when the tip contacts the surface, it has sufficient oscillation amplitude to overcome the tip-sample adhesion forces. Also, the surface material is not pulled sideways by lateral forces since the applied force is largely vertical.

(ii) **Phase Imaging:** An AFM operating in tapping mode can go beyond topographic imaging to detect spatial variation in surface composition, adhesion, friction, viscoelasticity and other properties by using a method that is termed as phase imaging. Phase imaging is the mapping of the phase lag between the periodic signal that drives the cantilever and the oscillations of the cantilever. Changes in the phase lag often indicate changes in the properties of the sample surface. Phase imaging is quite effective for contaminant identification, mapping of components in composite materials, differentiating regions of high and low surface adhesion or hardness and regions of different electrical or magnetic properties.

Sample properties can be explored by using the phase lag between the excitation (the signal that drives the cantilever to oscillate) and the response of the cantilever (the cantilever oscillation output signal) as an auxilary signal [118–120] because phase shift is directly correlated to the tip-sample interaction. Phase imaging is the spatial mapping of this phase lag. Changes in the phase lag often indicate changes in the mechanical properties of the sample surface. Simplified, the system can be modeled by an oscillator consisting of two damped springs representing the cantilever and the sample, respectively. The springs are each described by a characteristic spring constant and a damping constant. In the case of the sample these parameters basically model the viscoelastic interaction and are assumed to vary locally. The whole system can be described to a first approximation [121–124] by the external driving frequency ω_{ext}, the effective resonance frequency ω_{reso}, and an effective damping α_d which may vary locally. These effective quantities dominate the complete resonance behavior of the system, including the phase shift. Using the approximation in the resonant case [125] the phase shift ϕ is related to the effective parameters by the relation

$$tan\phi = 2\alpha_d \frac{\omega_{ext}}{\omega_{reso}^2 - \omega_{ext}^2} \qquad (3.7)$$

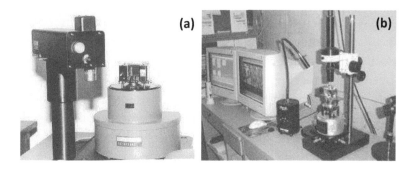

Figure 3.4: *Scanning probe microscopes (a) static and (b) dynamic modes.*

3.2.3 The Microscope

The controls for a typical scanning probe microscope are shown in Fig. 3.4 (a). The primary components are the probe head, probe cartridge, the manual XY translational stage and the scanner. The probe head contains the deflection sensor, which consists of a laser diode, a mirror and a position sensitive photodetector. The probe cartridge contains a removal cassette with a probe mounted on it. The probe cartridge slides in and out of the probe head on slide rails and clicks into place for precise positioning. The probe head itself slides on and off the XY translation stage. An electrical connector on the probe head plugs into a connector mounted on the back of the XY translation stage. The scanner consists of a piezo electric ceramic tube and position detectors which accurately measure the scanner tube position. The scanner is installed in an opening below the probe head, with the sample holder attached to the top of the scanner. The scanner rasters the sample beneath the probe tip during a scan. In another arrangement (Fig. 3.4 (b)) the scanner is attached to the sample holder. Two types of scanners can be used- $10 \ \mu m \times 10 \ \mu m$ with a vertical range of $2.5 \ \mu m$ and $0.4 \ \mu m \times 0.4 \ \mu m$ with a vertical range of $0.4 \ \mu m$. Measurements in the tapping mode can be carried out using this AFM, for which uncoated Si tip with a nominal tip radius of curvature of $5 - 10$ nm is used. The resonant frequency and spring constant of the cantilever which is attached to the tip for tapping mode measurements are $200 - 400$ kHz and $20 - 60$ N/m respectively.

3.3 Ellipsometry

Ellipsometry is a non-invasive, non-destructive measurement technique based on the change of polarization upon reflection from a sample. The optical properties, thickness of optically different layers, roughness and surface morphology can be obtained using this technique. The technique uses a relative change in polarization which can be measured accurately and reliably compared to the absolute values. This makes ellipsometric measurement very robust, sensitive, reliable and reproducible.

Like AFM, ellipsometry requires neither sample preparation nor special measurement environments. What is most interesting is the fact that this technique can provide information about layers that are thinner than the wavelength of the probing light and the depth resolution can be of the order of angstroms. Though the lateral resolution is limited by the optical wavelength and thus much worse than AFM, this is compensated by the absence of any upper cut-off on the sample size and, since the measurement technique does not mechanically disturb the sample, data on liquid surfaces can be taken without perturbing the surface. Thus this technique is widely used in various fields, from semiconductor physics to biophysics, to obtain dielectric properties and thickness of layers ranging from angstroms to micrometers. The major drawback of the technique, as we shall see, is the convolution of refractive indices and thickness in the data, which requires either an independent measurement of one of these quantities or an involved optical modeling.

3.3.1 Basic Principle

Ellipsometry measures relative change in polarization of reflected light parametrized by ellipsometric angles, Ψ and Δ. In general the polarization of an electromagnetic wave can be decomposed into two perpendicular components. The component which is parallel to the plane of incidence is called *p-polarized* and the component perpendicular to that plane is called *s-polarized* light. The ellipsometric angle Δ, which measures the change in phase of polarization, is defined as

$$\Delta = \Delta_p - \Delta_s \tag{3.8}$$

and the angle Ψ which related to the change in amplitude is defined as

$$\tan \Psi = \frac{(E_p/E_s)_{reflected}}{(E_p/E_s)_{incident}} \tag{3.9}$$

where E_p and E_s are respectively p and s components of the electric field. The above two relations can be put into a single relation as

$$\frac{r_p}{r_s} = \tan \Psi e^{i\Delta} \tag{3.10}$$

where r_p and r_s are complex reflection coefficients for p ans s components respectively.

3.3.2   Nulling Ellipsometry

The polarization of the incident light is changed upon reflection from the sample. It is possible that the reflected light is linearly polarized for some particular elliptic polarization of the incident light. We can change the elliptic polarization of the incident light by changing *polarizer* and *compensator* angles. If the reflected light is linearly polarized it is possible to extinguish the intensity by suitably adjusting the analyzer angle (A). At this nulling condition the signal of the detector is minimum. For the nulling condition, the ellipsometric angles (Δ, Ψ) and the polarizer (P), analyzer (A) and compensator (C) angles are related by

$$r_p/r_s = \frac{\tau_c \tan(P - C) + \tan C}{\tan(P - C)\tan(C - 1)\tan A} \tag{3.11}$$

where $\tau_c = \tan \Psi_c e^{i\Delta_c}$ is the complex transimittance ratio of the compensator. The results are significantly simplified for $\tau_c = i$ i.e., $C = \pi/4$. The minimum detector intensity obtained for two set of polarizer and analyzer angles P_1, A_1 and P_2, A_2 related by $P_2 = +/- P_1$, $A_2 = -A_1$. In addition, the ellipsometric angles can be obtained from the simple relations $\Psi = |A_1|$, $\Delta = 2P_1 + \pi/2$.

Nulling ellipsometry gives the values of Δ and Ψ as functions of wavelength (called *spectroscopic or multi-wavelength ellipsometry*) and/or angle of incidence (for single-wavelength ellipsometry only angle of incidence can be varied). The complex refractive index and thickness of a sample are usually related to ellipsometric angles by complicated relations. It is practially impossible to solve for thickness or refractive index of a sample from those relations. To obtain the optical properties or thickness from the ellipsometric angles one need to use *optical modeling*, where data aquired in the ellipsometry measurement are compared to a model to obtain the physical properties of the sample. In the optical model, the sample is usually assumed to be composed of different discrete, well-defined, optically homogeneous and isotropic layers and fit the values of Δ and Ψ obtained from this model to the experimental values by an iterative method. The thickness and/or the refractive index are then extracted from the fit parameters.

3.3.3 Imaging Ellipsometry

In imaging ellipsometry , nulling ellipsomety is done with a spatially resolving detector to obtain ellipsometric high contrast image from the surface. Light from the illuminated area of the sample is made incident onto the detector camera. The spatially resolving camera can distinguish the signal coming from different parts of the sample. For a particular set of P, C and A values some part of the areas will satisfy the nulling condition with minimum detector signal and other parts will appear brighter with higher intensities. Changing the P, C and A values one can find the null for the entire illuminated region. In this way, one can have a two-dimensional map of ellipsometric angles that can be transformed into thickness map of the sample using optical modelling.

3.3.4 Brewster Angle Microscopy

Brewster angle microscopy is a special case of ellipsometry where the p-polarized light is incident on the film on a substrate at the *Brewster angle* of the air/substrate media, which allows only the

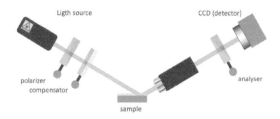

Figure 3.5: *Setup for measuring ellipsometric angles in PCSA configuration.*

s-polarized light to be reflected, hence in the absence of any film the image should be black as no light is received by the spatially resolving detector camera. Any film on the substrate results in reflection of incident light. Thus a high contrast image of lateral morphology of a film can be obtained using this technique. This technique is widely used to see the two-dimensional phases of Langmuir monolayers on water surface and to capture various phase transition and relaxtion dynamics of Langmuir monolayers.

3.3.5 The Ellipsometer

The optical components and the setup for ellipsometry are shown in Fig. 3.5 while an imaging ellipsometer with Brewster angle microscopy facility is shown in Fig. 3.6.

Figure 3.6: *Imaging ellipsometer.*

A laser is used as a light source to produce unpolarised parallel light. The light becomes linearly polarized after it passes through the polarizer. The direction of polarization can be set to any value by rotating the polarizer. The polarized light is then allowed to pass through the compensator which acts as an optical retarder. The compensator transforms the linearly polarized light into an elliptically polarized light and also the ellipticity of the transformed light can be changed by rotating the compensator. After reflection from the sample the polarization of the light is changed depending on the dielectric properties and thickness of the sample. Finally the state of polarization of the reflected light is analyzed with analyzer. From the polarizer, compensator and analyzer angles Δ and Ψ can be obtained.

3.4 X-Ray Reflectivity

X-Ray reflectivity (XRR) has emerged as a very useful tool for the investigation of the structural properties of ultrathin films. This technique provides excellent spatial resolution, down to $\sim$ 0.5 nm with penetration depths over hundreds of nanometers. It probes the variation in electron density of the sample. It is non-destructive and, consequently, repetitive measurements on one specimen can be performed. The specimen does not require special conditioning and measurements can be done under ambient conditions. Moreover, using x-rays in grazing incidence increases the sensitivity to the surface [50,126,127]. The non-destructive nature of x-ray scattering technique and its ability to explore structural properties of crystalline and non-crystalline solids [50, 126–129], liquids [130–133], thin films and their buried interfaces [134–136] makes this technique an indispensable probe for structure determination.

3.4.1 Basic Principle

Fig. 3.7 shows the scattering geometry in the general case of a surface experiment. The plane of incidence contains the incident

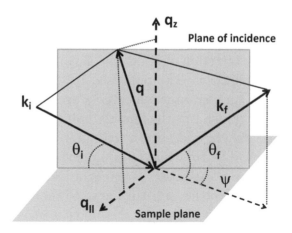

Figure 3.7: *Schematic showing a generalized scattering geometry.*

wave-vector $\mathbf{k}_i$ and the normal to the surface. If $\mathbf{k}_f$ is the scattered wave vector, then the momentum transfer vector $\mathbf{q}$ is given by

$$\mathbf{q} = \mathbf{k}_f - \mathbf{k}_i \qquad (3.12)$$

The components of $\mathbf{q}$ in x, y and z directions are given by

$$
\begin{aligned}
q_x &= k_0(\cos\theta_f \cos\psi - \cos\theta_i) & (3.13) \\
q_y &= k_0(\cos\theta_f \sin\psi) & (3.14) \\
q_z &= k_0(\sin\theta_f + \sin\theta_i) & (3.15)
\end{aligned}
$$

where $k_0 = 2\pi/\lambda$, λ being the wavelength of the incident radiation. In a reflectivity experiment, we work in the plane of incidence and thus have the in-plane angle $\psi = 0$. In specular reflectivity, incident angle (θ_i) and the scattered angle (θ_f) are kept equal. This gives information about the thickness and electron density profile of the individual layers and the interfacial roughness [50, 126, 135, 137]. Off-specular diffuse scattering ($\theta_i \neq \theta_f$) provides better understanding about the in-plane correlation and also the correlation between the interfaces [50, 132, 138]. In case of grazing incidence diffraction (GID) experiments, one varies the in-plane angle ψ. This gives in-plane lattice structure of the sample [27, 139, 140].

In general, in x-ray reflectivity measurements, a well collimated monochromatic x-ray beam is allowed to fall on the sample surface at a grazing angle θ_i (starting from few milliradians) and the reflected intensity is recorded in the plane of incidence at an angle θ_f. If the angle of incidence of impinging x-rays is sufficiently small, the penetration depth and scattering is limited to the near surface region. Generally reflectivity data is taken at angles considerably larger than the critical angle of total external reflection and therefore the penetration depth is of the order of thousands of angstroms. The sensitivity is then obtained by interference of x-rays scattered from the different layers having difference in their electron densities at different depths in the sample.

3.4.2 Formalism

An electromagnetic plane wave given by its electric field vector $\Psi(\mathbf{r}) = \Psi_0 \exp(i\mathbf{k_i}.\mathbf{r})$, which penetrates into a medium characterized by an index of refraction $n_{ri}(\mathbf{r})$, propagates according to the Helmholtz equation

$$(\nabla^2 + k_0^2 n_{ri}^2)\Psi(\mathbf{r}) = 0 \qquad (3.16)$$

where $k_0 = 2\pi/\lambda$ is the modulus of the wave vector $\mathbf{k_i}$ and λ denotes the wavelength of the radiation. In general, the index of refraction for an arrangement of N atoms per unit volume, which may be assumed to be harmonic oscillators with resonance frequency ω_j for the jth atom, is expressed as [141]

$$n_{ri}^2 = 1 + \left(\frac{Ne^2}{\varepsilon_0 m}\right) \sum_{j=1}^{N} \frac{f_j}{\omega_j^2 - \omega_{em}^2 - 2i\omega_{em}\eta_j} \qquad (3.17)$$

where ω_{em} is the frequency of the incoming electromagnetic wave, e is the charge and m is the mass, respectively, of the electron, the η_j are damping factors, and the f_j denote the oscillator strength of the jth atom, representing the collective oscillations of the band electrons. It should be noted that in general f_j are complex numbers, $f_j = f_j^0 + f_j'(E) + if_j''(E)$, where $f_j'(E)$ and $f_j''(E)$ take into account dispersion and absorption corrections depending on the radiation energy E [142]. For x-rays $\omega_{em} > \omega_j$, and Eqn. 3.17 may be replaced by [143–145]

$$n_{ri} = 1 - \delta + i\beta \tag{3.18}$$

with the dispersion and absorption terms

$$\delta = \frac{r_0\lambda^2\rho_{el}}{2\pi} \sum_{j=1}^{N} \frac{f_j^0 + f_j'}{Z} = \frac{r_0\lambda^2}{2\pi} N_A \sum_{j=1}^{N} \frac{\rho_j}{A_i}(Z_j + f') \tag{3.19}$$

and

$$\beta = \frac{r_0\lambda^2\rho_{el}}{2\pi} \sum_{j=1}^{N} \frac{f_j''}{Z} = \frac{r_0\lambda^2}{2\pi} N_A \sum_{j=1}^{N} \frac{\rho_j}{A_j} f_j'' \tag{3.20}$$

It should be emphasized that δ is always positive. In Eqns. 3.19 and 3.20 we have introduced: (1) the classical electron radius $r_0 = e^2/(4\pi\varepsilon_0 mc^2) = 2.814\times10^{-5}$ Å (2) the total number of electrons $Z = \sum_j Z_j$, where Z_j denotes the number of electrons in j^{th} atom and (3) the electron density ρ_{el} as a function of the spatial co-ordinates $r = r(x, y, z) = r(r, z)$. The quantities f_j^0 are **q** dependent, where $\mathbf{q} = \mathbf{k_f} - \mathbf{k_i}$ is the wave vector transfer (k_i, k_f are the wave vectors of the incident and scattered x-ray plane waves). This has to be taken into account when measurements over a large q region are analyzed [90]. However in the region of grazing incidence and exit angles, θ_i and θ_f, respectively, the wave vector transfer is small, and f_j^0 may be approximated with high accuracy by $f_j^0 \approx Z_j$. N_A is Avogadro's number, ρ_j is the density of element j with atomic weight A_j [40, 76]. For x-rays, $\delta \sim 10^{-6}$ and β is usually one or two orders of magnitude smaller. In case of a homogeneous medium and far away from absorption edges, Eqns. 3.19 and 3.20 can be rewritten in terms of the electron density ρ_{el} and the linear absorption coeffcient (or mass absorption coeffcient), μ as

$$\delta = \lambda^2 \rho_{el} r_0/2\pi \tag{3.21}$$

and

$$\beta = \mu\lambda/4\pi \tag{3.22}$$

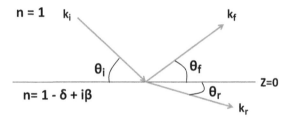

Figure 3.8: *Schematic showing incident, scattered, and refracted beams from an interface separating two media.*

In vacuum, the z component of the momentum transfer vector, normal to the surface or x-y plane, is given by

$$q_{z,0} = \frac{2\pi}{\lambda}(sin\theta_i + sin\theta_f) \tag{3.23}$$

In specular condition $\theta_i = \theta_f$, we can write,

$$k_{z,0} = \frac{q_{z,0}}{2} = 2\pi sin\theta_i/\lambda \tag{3.24}$$

For a single vacuum/medium interface, Snell's law of refraction gives

$$cos\theta_i = (1 - \delta)cos\theta_r \tag{3.25}$$

where θ_r is the exit angle of the refracted radiation as shown in Fig. 3.8. Thus if $\theta_r = 0$, and since δ is very small, the critical angle is

$$\theta_c \approx \sqrt{2\delta} = \lambda\sqrt{r_0\rho_{el}/\pi} \tag{3.26}$$

Now we consider a perfectly smooth surface of a medium on which x-ray hits at a grazing angle θ_i. For incident angles $\theta_i \leq \theta_c$ the phenomenon of total external reflection occurs. The x-ray does not penetrate far into the medium and all incoming radiation is reflected. When $\theta_i > \theta_c$, the incident beam reflected and transmitted at angles θ_f and θ_r respectively. In specular condition $\theta_i = \theta_f$ and only the z-component of momentum transfer vector **q** will survive (i.e., $q_x = q_y = 0$). The critical value of momentum transfer vector q_c corresponding to θ_c can be written as

$$q_c^2 = \left(\frac{4\pi}{\lambda}\right)^2 \theta_c^2 = 16\pi \rho_{el} r_0 \tag{3.27}$$

The z-component of momentum transfer vector in the medium $q_{z,1}$ can be written as

$$q_{z,1} = (q_{z,0}^2 - q_{c,1}^2)^{1/2} \tag{3.28}$$

Here we have used the suffix "0" and "1" for air (or vacuum) and medium respectively. By using the appropriate boundary condition for the electric field and its derivative at the air/medium interface one can obtain the Fresnel formulas for reflection and transmission coefficient as [50, 134–137]

$$r_{01} = \frac{k_{z,0} - k_{z,1}}{k_{z,0} + k_{z,1}} = \frac{q_{z,0} - q_{z,1}}{q_{z,0} + q_{z,1}} \tag{3.29}$$

and

$$t_{01} = \frac{2q_{z,0}}{q_{z,0} + q_{z,1}} \tag{3.30}$$

The specular reflectivity can then be written as

$$R = r_{01} \cdot r_{01}^* \tag{3.31}$$

R_F is simply the Fresnel reflectivity and can be written in terms of q_z as

$$R_F(q_{z,0}) = \left| \frac{1 - [1 - (q_{c,1}/q_{z,0})^2]^{1/2}}{1 + [1 - (q_{c,1}/q_{z,0})^2]^{1/2}} \right|^2 \tag{3.32}$$

This is shown in Fig. 3.9 for the interface between silicon and air. For $q_{z,0} > 3q_{c,1}$, the Fresnel reflectivity may be well approximated by

$$R_F(q_{z,0}) \propto (q_{c,1}/q_{z,0})^4 \tag{3.33}$$

Therefore for an infinitely sharp interface the reflectivity will vary with $q_{z,0}^{-4}$ at high values of $q_{z,0}$. In reality, interfaces are not smooth and the effect of roughness has to be taken into account, during data processing. This is discussed in Chapter 4.

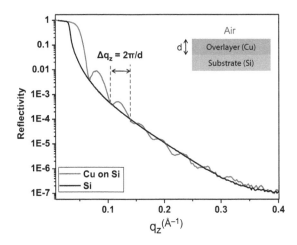

Figure 3.9: *A typical reflectivity profile of Cu deposited on Si, schematic of which is shown in inset.*

3.4.3   Grazing Incidence X-Ray Diffractometer

Any laboratory x-ray diffractometer consists of three main parts: (a) the x-ray source and associated beam optics, (b) the goniometer, and (c) the detector and associated electronics (Fig. 3.10).

(a) X-Ray source and beam optics

In a typical x-ray diffractometer, a sealed tube of either Cu or Mo generates x-rays of the required radiation energy, focal spot and intensity. The focal spot (also called focal spot on target) and takeoff angle are critical features in the production of x-rays by sealed tube. Sealed tubes, of course, produce x-rays by bombarding the target sample with electrons generated from the filament (cathode). The area bombarded by electrons is called focal spot on target and the angle between the primary x-ray beam and the surface is called takeoff angle. Sealed tubes normally have 2 to 4 beryllium windows through which x-rays exit. The focal spot is typically rectangular with a length-to-width ratio of 10:1. The projection along the length of the focal spot at a takeoff angle from the anode surface is called spot focus. The takeoff angle can be set

Figure 3.10: *Grazing incidence x-ray diffractometer.*

from 3° to 7° (6° for most systems). For fine focus the focal spot size at anode is 0.4×8 mm^2 and spot focus size is 0.4×0.8 mm^2. The emerging primary x-rays then pass through the beam optics components consisting of a monochromator, a pinhole collimator and cross-coupled Göbel mirrors. A graphite crystal monochromator is typically used which gives the strongest beam intensity. This monochromator cannot resolve the $K_{\alpha 1}$ and $K_{\alpha 2}$ lines, so it is aligned to K_α line (wavelength 1.541838 Å). X-rays of other wavelengths are filtered out by the monochromator. The monochromatized x-ray beam then passes through a pinhole collimator which is normally used with a monochromator or a set of cross-coupled Göbel mirrors. The Göbel mirror is a parabolic multilayer mirror and it yields an intense and parallel beam. In specular reflectivity geometry, χ and ϕ are set to 0 and the horizontal $\theta - 2\theta$ mode is used.

(b) Goniometers and sample stages

The goniometer in a typical grazing incidence diffractometer is a high-precision, two-circle goniometer with independent stepper motors and optical encoders for θ and 2θ circles. The goniometer can be used in horizontal θ-2θ, vertical θ-2θ and vertical θ-θ geometries. The sample stages are usually mounted on the inner

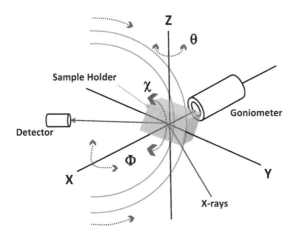

Figure 3.11: *Rotation geometry of the θ, ϕ and χ motors in VXRD setup with 1/4-circle Eulerian cradle.*

θ circle of the goniometer. In θ-2θ mode, the sample rotation is defined as ω rotation, so a sample stage directly mounted on the goniometer inner circle is also called ω-stage. The used sample stages are basically fixed-χ, two position, XYZ and 1/4-circle Eulerian cradle. This cradle has two rotational (χ and ϕ) and three translational (X, Y and Z) motions. The corresponding θ, ϕ and χ motors' movements have been shown in Fig. 3.11.

(c) Detector

In most cases, the scattered beam is detected using NaI scintillation (*point*) detector. The detector interface board generates the high-voltage required for the operation of scintillation and proportional counters. It contains a high speed pulse amplifier with a pulse shaping stage, line shift correction and baseline restoration unit and also two discriminator windows including complete pulse counting unit. In addition there is a digital count input for various kinds of detectors having a digital pulse output.

3.5   Infrared Spectroscopy

Infrared radiation is another name for heat. All objects in the universe at the temperature above absolute zero give off infrared radiation. When infrared radiation interacts with matter it can be absorbed, causing the chemical bonds in the material to vibrate. Chemical structural fragments within molecules, known as *functional groups*, tend to absorb infrared radiation in the same wavenumber range regardless of the structure of the rest of the molecule. This allows proper identification of functional groups and hence the molecular structure by infrared (IR) spectroscopy. The theory of vibrational spectroscopy and the basis of molecular structure determination using IR spectroscopy have been discussed in Chapter 4. In general, IR radiation is allowed to fall on the sample whose structure is to be studied, and either the transmitted or the reflected light is collected by a suitable detector. A plot of the intensity measured by the detector versus the wavenumber is called an infrared spectrum and it gives information about the characteristic frequencies of the sample.

There are two distinct types of infrared spectrometers in common use: *Dispersive instruments* and *Fourier transform instruments* . In the dispersive instruments the radiation is physically split up into its constituent wavelengths either before or more usually after, it passes through the sample and the different wavelengths are processed in sequence. In the Fourier transform instrument radiation of all wavelengths passes through the sample to the detector simultaneously. The detector measures the total transmitted intensity as a function of the displacement of one of the mirrors in a double beam interferometer, usually of the Michelson type, and the separation of the different wavelength is subsequently done through a Fourier transform of the intensity pattern. Fourier transform instruments are faster, have higher signal-to-noise ratios and hence provide better resolution. The Fourier Transform Infrared (FTIR) spectrometer is extensively used in soft matter research. Its working principle and basic modes of operation are described below.

3.5.1 Basic Principles of Fourier Transform Spectroscopy

Inside an FTIR spectrometer, a broadband IR source generates the beam, which passes through a Michelson interferometer Fourier transform instruments [146]. The purpose of an interferometer is to take a beam of light, split it into two beams, and make one of the light beam travel a different optical path length than the other. The difference in path length traveled by these two light beams is called the optical path difference (δ_{pd}).

The schematic of the FTIR spectrometer is shown in Fig. 3.12. At the intersection of the four arms of the interferrometer is a beam splitter, which is designed to transmit half the radiation that impinges upon it, and reflect half of it. Movement of mirror M_1 causes a change in the optical path difference between two beams of monochromatic radiation of wavelength λ, that pass through the interferometer. Constructive interference occurs whenever $\delta_{pd} = n$ λ, where n is an integer, such that the intensity of the recombined beams, as collected by the detector, is maximum. Again, when δ_{pd} = $(n+1/2)$ λ, the beams interfere destructively, and gives zero intensity. For intermediate values of path difference, a combination of constructive and destructive interference takes place, and the intensity is somewhere between very bright and very weak. The intensity measured by the detector varies cosinusoidally with the

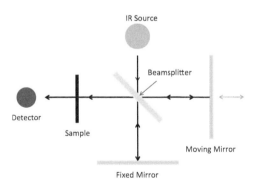

Figure 3.12: *Schematic of a Michelson interferometer.*

displacement of the mirror. For monochromatic light of wavenumber $\bar{\nu}(= 1/\lambda)$ and path difference δ_{pd}, the measured intensity, $I(\bar{\nu})$, is related to the incident intensity, $I_0(\bar{\nu})$, by the equation

$$I(\delta_{pd}) = \frac{1}{2}I_0(\bar{\nu})T(\bar{\nu})(1 + \cos 2\pi\delta_{pd}\bar{\nu}) \qquad (3.34)$$

where $T(\bar{\nu})$ is the transmittance of the sample at $\bar{\nu}$. The path difference is equal to twice the mirror displacement, measured from the position where the two paths are equal.

When polychromatic radiation passes through the interferometer, as it does when the interferometer is used for recording infrared spectra, the intensity measured for any value of δ_{pd} is the integral of the above expression over the range of wavenumber involved. The intensity expressed as a function of δ_{pd}, apart from a constant added term $\frac{1}{2} \int I_0(\bar{\nu})T(\bar{\nu})d\bar{\nu}$, is proportional to the Fourier transform $\int I_0(\bar{\nu})T(\bar{\nu}) \cos(2\pi\delta_{pd}\bar{\nu})d\bar{\nu}$, of the spectrum $I_0(\bar{\nu})T(\bar{\nu})$ and this is called the *interferogram*. The detector signal and corresponding mirror displacement are fed to an on-line computer which calculates the inverse Fourier transform to provide the intensity as a function $I(\bar{\nu})$ of the wavenumber, i.e., a spectrum in the conventional form. The effect of a double-beam instrument (useful to eliminate the background effect) is obtained by recording an additional interferogram corresponding to $I_0(\bar{\nu})$ after removing the sample to provide a new path from the interferometer to the detector which does not pass through the sample but is otherwise equivalent to the original path. In both methods the transmittance or absorbance of the sample is obtained by calculation from the stored spectra.

Modes of Operation

An FTIR spectrometer operates in two distinct modes - *Transmission* mode and *Reflectance* mode, the latter including the *Attenuated Total Reflectance* (ATR), *Specular Reflectance* and *Diffuse Reflectance* modes. Depending on the sample type, appropriate mode of operation (also known as sampling technique) is chosen.

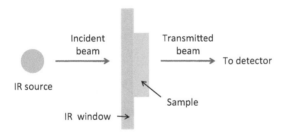

Figure 3.13: *Schematic of the transmittance mode of FTIR oper-ation.*

(a) Transmission mode

It is an universal technique (i.e., it works for solids, liquids and gases) and also the most popular method of obtaining an infrared spectra. Here, the IR beam is directly passed through the sample as shown in Fig. 3.13. The IR spectra thus records the informa-tion of the bulk sample. The main advantage of this technique is that the spectra has high S/N ratio. However, in transmis-sion sampling, the major disadvantage is the limitation in sample thickness. Samples that are thicker than $\sim$ 20 microns absorb too much infrared radiation making it impossible to obtain a spec-trum. Typically, solid samples are investigated as KBr pellets or mulls to increase the transmittance and as thin films; liquids and gases are analysed using demountable sealed cells with IR trans-parent windows. In our research, we have used this mode for samples deposited on substrates that are nearly or completely IR transparent.

(b) Reflectance mode

In Infrared Reflection Absorption Spectroscopy (IRAS) or reflectance mode, the IR beam is bounced off the surface instead of passing through it (Fig. 3.14). This sampling technique is surface sensitive, as the IR beam penetrates very little inside the sample ($\sim \mu m$), and hence there is no "thickness problem" as in trans-mission mode. However, reflectance data have low S/N ratio and

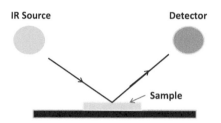

Figure 3.14: *Schematic of the reflectance mode of operation.*

also more useful for reflecting surfaces such as thin films on reflecting substrates. Inspite of its disadvantages, reflectance mode finds applicability for its surface sensitivity. There are three distinct modes of operation under reflectance sampling technique. These are discussed below.

(i) Attenuated Total Reflectance (ATR): This mode is used for solid, liquid, semi-solid samples as well as thin films. Majority of our FTIR data have been taken in this mode, to eliminate problem of absorption due to substrate. The schematic of an ATR accessory is shown in Fig. 3.15. At the heart of the accessory is a crystal which is IR transparent (typically Zinc Selenide) and has high refractive index. The IR beam is incident on the crystal surface at an angle such that the radiation undergoes total internal reflection and hence light is reflected off the crystal surface. In-

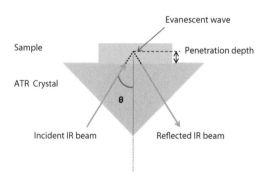

Figure 3.15: *Schematic of attenuated reflectance mode of operation.*

side the crystal, the IR beam produces an evanescent wave that penetrates a small distance beyond the crystal surface. When the sample is placed on the crystal at zero contact angle, the evanescent wave is attenuated by the sample's absorbance. The depth of penetration DP is defined as the depth at which the evanescent wave is attenuated to $(1/e)$ of its total intensity and is given by

$$DP = 1/(2\pi\bar{\nu}N_C(sin^2\Theta - N_{SC}^2)^{1/2}) \qquad (3.35)$$

where $\bar{\nu}$ = wavenumber, N_C = crystal (ZnSe) refractive index, Θ = angle of incidence and $N_{SC} = N_S/N_C$, N_S being the sample refractive index.

(ii) Specular Reflectance: This mode is used for samples that have a smooth surface. However, for investigating thin films or coatings under specular condition, reflection-absorption mode is used. For this, the thin film or coating is deposited on a highly polished surface such as silicon, such that the IR beam passes through the sample and gets reflected from the polished surface and passes through the sample once more. In the process, the sample absorbs part of the IR radiation. By comparing the obtained spectra with that of the specular reflectance spectra of the polished surface, the reflection-absorption spectra of the thin film or coating is obtained. Spectra of Langmuir monolayers can be taken by attaching a mini LB trough to this set-up in place of the polished plate.

(iii) Diffuse Reflectance: Diffuse Reflectance Infrared Fourier Transform Spectroscopy (DRIFTS) is used to obtain the IR spectra of powders and other solid materials. Here, incoming radiation is directed onto a spherical or ellipsoidal focusing mirror and consequently focused on the powdered sample which is mixed with KBr. The diffusely reflected radiation is collected by a second spherical or ellipsoidal mirror and eventually focused onto the detector. The obtained spectra is compared with the DRIFT spectra of pure KBr to give the DRIFT spectra of the sample.

Figure 3.16: *Fourier transform infrared spectrometer.*

3.5.2   The Spectrometer

The FTIR Spectrometer (Fig. 3.16) generally has a working range from 10000 cm^{-1} to 370 cm^{-1} (Near to Mid IR) and 700 cm^{-1} to 30 cm^{-1} (Far IR). A voltage-stabilized, air-cooled tungsten halogen lamp (for Near and Mid IR) and wire coil at 1350 K (Far IR) act as IR source. The interferrometer is of Michelson type, with a KBr beamsplitter for Mid IR and grid for Far IR regions. The detectors are generally Deuterated TriGlycine Sulphate (DTGS) with KBr window and Mercury Cadmium Telluride (MCT) for the Near and Mid IR and DTGS with polythene window for Far IR. These instruments can be operated in (a) transmission, (b) attenuated total reflectance (ATR) and (c) specular reflectance modes (Fig. 3.17) using ZnSe crystals as substrates for ATR and Au-coated mirrors for the specular reflectance modes with suitable holders and sample cells. For taking data on Langmuir monolayers a small PTFE trough can be attached onto the Reflectance set up. The interferometer resolution range may vary from 0.2 cm^{-1} to 64 cm^{-1}. The optical path difference velocity, which gives the scan speed, varies from 0.05 cm/sec to 5.0 cm/sec. A HeNe laser ($\bar{\nu} = 15798.01$ cm^{-1}) is generally used as the wavelength standard.

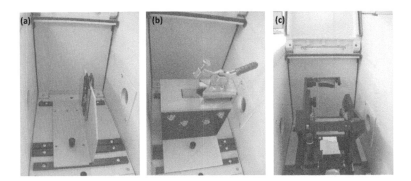

Figure 3.17: *The three stages for FTIR sampling (a) transmittance, (b) attenuated total reflectance, and (c) specular reflectance.*

3.6 Near Edge X-Ray Absorption Fine Structure Spectroscopy

Near edge x-ray absorption fine structure (NEXAFS) spectroscopy is a method to characterize materials by probing the unoccupied electronic states. NEXAFS is also called x-ray absorption near edge spectroscopy or XANES. Today, the term NEXAFS is typically used for soft x-ray absorption spectra and XANES for hard x-ray spectra. NEXAFS just needs a monochromatically tunable light source so that such measurements can be performed at any synchrotron radiation source of adequate energy. Due to sharp transitions in x-ray spectra NEXAFS is one of the favorite experimental techniques to study organic thin films [147].

3.6.1 Basic Principles

The intensity of x-ray transmitted through a sample decreases due to x-ray absorption. This general absorption due to various mechanisms like inelastic scattering processes (Compton effect, photoelectron and Auger electrons production, etc.) is generally small and it reduces with reduction in x-ray wavelength. However, when the photon energy corresponds to the binding energy of a core electron, the x-ray absorption cross section increases abruptly, and

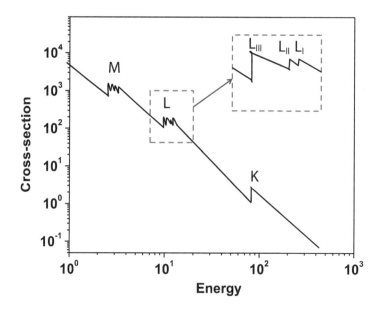

Figure 3.18: *Schematic of a NEXAFS spectrum.*

then decreases monotonically above the core edge. Thus an edge-like structure is observed in the absorption spectrum. Based on the origin of the excited electron, the absorption edge can be classified as K-edge (corresponding to the excitation of a 1s electron to make it just free, i.e., just beyond the ionization potential), L-edge (L_I-, L_{II}-, and L_{III}-edge corresponds to the excitation of a 2s, $2p_{1/2}$ and $2p_{3/2}$ electron, respectively), and so on. Fig. 3.18 presents a typical x-ray absorption cross section as a function of x-ray energy [148]. With the increase of photon energy, the x-ray absorption cross section that describes the probability of absorption decreases.

X-ray absorption cross-section: The x-ray absorption cross section and optical oscillator strength are the two main factors to describe the absorption probabilities. The absorption cross section is defined as the number of electrons excited per unit time divided by the number of incident photons per unit time per unit area [147]. Within the dipole approximation the x-ray absorption cross section is given by

$$\sigma_x = \frac{4\pi^2 \hbar^2}{m^2} \frac{e^2}{\hbar c} \frac{1/\hbar}{\nu_p} < \Psi_f \mid \mathbf{E} \bullet \mu_\mathbf{D} \mid \Psi_i >^2 \rho_f(\mathbf{E}) \qquad (3.36)$$

where σ_x is the x-ray absorption cross section in cm^2, which is usually expressed in barn (1 cm$^2 = 10^{24}$ barn (b)), $\hbar = h/2\pi$, h is the Planck constant, c is the speed of light, e is the electron charge, m is the mass of the electron, ν_p is the photon frequency, $\mathbf{E}$ is the electric field vector of x-ray photons, $\mu_\mathbf{D}$ is the dipole operator, ψ_f is the wavefunction of the final state, ψ_i is the wavefunction of the initial state, and $\rho_f(\mathbf{E})$ is the energy density of the final state.

Eq. 3.36 is the basis of theoretical calculations of x-ray absorption spectra. The outcome of these theoretical calculations is often expressed as the optical oscillator strength f_{op}, which is related to the x-ray absorption cross section by

$$\sigma_x(\mathbf{E}) = C_0 \frac{df_{op}}{dE} \qquad (3.37)$$

where $C_0 = 2\pi^2 e^2 \hbar / mc = 1.1 \times 10^2$ Mb eV. This oscillator strength f_{op} is the energy integral of the x-ray absorption cross section, and therefore, a measure of the intensity of a resonance. The optical oscillator strength is often also called the *"f number"* of the transition

$$f_{op} = \frac{2}{m\hbar^2 \nu_p} < \Psi_f \mid \mathbf{E} \bullet \mu_\mathbf{D} \mid \Psi_i >^2 \qquad (3.38)$$

The calculated oscillator strengths f_{op} for transitions can be converted to x-ray absorption cross section using Eq. 3.39 [147]

$$\sigma_x = \frac{2\pi^2 e^2 \hbar}{mc} f_{op} \rho_f(E) \qquad (3.39)$$

X-ray absorption spectrum: Generally, the X-Ray absorption spectrum can be divided into two parts, the Near Edge X-Ray Absorption Fine Structure (NEXAFS) that corresponds to the absorption up to ~50 eV above the absorption edge and Extended X-Ray Absorption Fine Structure (EXAFS), which corresponds to the absorption up to 1000 eV above the absorption

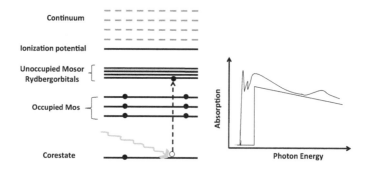

Figure 3.19: *Schematic of a NEXAFS spectrum.*

edge. EXAFS spectroscopy is used mostly in the study of inorganic species and ordered systems for determining the numbers, types, and distances of the backscattering atoms surrounding the absorbing atoms [149]. However, core edges in low Z atoms are too close together and for disordered materials the cross-section at high k are too small to permit EXAFS analysis. For organic molecular species, the NEXAFS part of the spectrum is the most useful.

Principles of NEXAFS Spectroscopy

Fig. 3.19 outlines the principles of NEXAFS spectroscopy. Near the K-shell absorption threshold, a series of fine structures are superimposed on the absorption edge. In organic molecules, these fine structures are dominated by resonances arising from transitions of 1s core electrons to unoccupied π^* or σ^* orbitals, depending on the covalent bonding in the molecule, as well as to Rydberg orbitals (left side of Fig. 3.19). The characteristic features of the K-shell spectrum are shown in the right part of this Fig. 3.19. NEXAFS spectrum is usually dominated by two types of resonances, with the ionization potential (IP) or k-edge as a boundary. Resonances below the IP correspond to the excitation of a core electron to a bound orbital. These orbitals are usually of π^* or Rydberg character and sometimes of σ^* character for saturated species, such as n-alkanes. These resonances are usually sharp and well defined. Resonances above the edge usually correspond to the excitation of a core electron to an unbound orbital of σ^* character, as well as double excitation. These resonances are usually broad. At the IP, a step-like increase is observed, from the onset of the core edge (Fig. 3.18). NEXAFS spectroscopy can be used for qualitative analysis of organic materials to obtain information like which element and what functional groups are present in the sample. This spectroscopy can also be used to quantitatively determine the chemical composition of a complex molecule. Furthermore, polarization dependent NEXAFS spectroscopy allows orientation analysis. We have conducted polarization dependent NEXAFS analysis for our research.

3.6.2 Synchrotron X-Ray Sources

Due to the nature of the technique, NEXAFS measurements require exceptionally bright x-ray sources of *tunable energy*. Such x-rays are available from *synchrotron sources* and before going further, we will briefly discuss about synchrotron radiation.

Synchrotron Radiation

A kind of light called "synchrotron radiation" was accidentally discovered at the electron synchrotron of the General Electric Company, USA, in 1947. Synchrotron radiation refers to a continuous band of electromagnetic spectrum including infrared, visible light, ultraviolet, and x-rays regions. When electrons or positrons are subjected to an acceleration perpendicular to their velocity, they begin to travel in a circular path and emit electromagnetic radiation. If kinetic energy of electrons or positrons is low as compared to the rest mass energy given by mc^2, i.e., when the electron velocity is non-relativistic, the angular distribution of the emission is extended over a large range of angles. However, when their speed approaches the speed of light c, the amount of energy radiated increases dramatically and relativistic effects cause the pattern to be folded into a sharp forward cone with a very small opening angle given by ω_{OA}^{-1}, where $\omega_{OA} = (1 - v^2/c^2)^{-1/2}$, v being particle velocity.

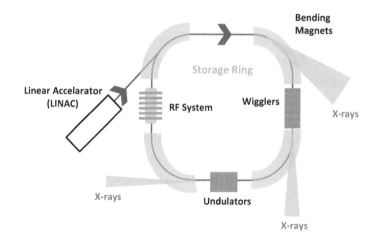

Figure 3.20: *Schematic of a synchrotron source.*

Many synchrotron centers over the world have been built to produce synchrotron radiation. An example is the ELETTRA synchrotron in Trieste, Italy, which is a 2.4 GeV, third-generation synchrotron radiation source. The linear accelerator (LINAC) speeds up the electrons or positrons to 99.999995% of the light velocity and injects them with energy of 1-2 GeV into the storage ring, where they circulate in UHV pipes for several hours (Fig. 3.20). The storage ring is roughly hexagonal in shape and has a circumference of about 100 meters. A series of magnets mounted at the ring steers the electrons along circular arcs and synchrotron radiation is continuously emitted tangentially to the arcs.

By using superconducting permanent magnets in periodic arrays (wigglers) the trajectory of charged particles can be bent strongly causing forceful local acceleration, resulting in high brilliance of synchrotron radiation. Periodic magnetic structure of undulators built into straight sections of a storage ring leads to oscillating motion of electrons in a way that they can be considered as oscillators with a particular natural frequency (main undulator harmonic) primarily defined by the strength of the magnetic field and the period of undulators. Anharmonicity of undulators results in presence of higher harmonics in the spectrum. In these insertion devices, i.e., undulators, spectral function of radiation is changed from continuous spectrum characteristic of bending magnets to a series of sharp peaks corresponding to these harmonics. Such spectral redistribution leads to tremendous increase of photon flux within the undulator harmonics at the cost of signals from other spectral regions. Therefore, the undulators are especially suited for experiments like angle-resolved photoemission (ARPES), where photon beams of high intensity, but fixed photon energy should be used.

In the case of NEXAFS studies, on the other hand, where tunable light in wide ranges of photon energies is required, preferably radiation from bending magnets has to be applied. In all cases the emitted light is channeled through beamlines to the experimental stations where experiments are conducted. The main attractive properties of synchrotron radiation are: (1) high intensity; (2) continuous spectrum from bending magnets and wigglers, while a series of harmonics from undulators; (3) excellent collimation; (4) low emittance; (5) pulsed time-structure, and (6) polarization. As an example, the intensity of synchrotron x-rays is more than million times higher than that of x-rays from conventional x-ray tubes.

Synchrotron radiation can be obtained over a very broad range of photon energies, from microwaves to hard x-rays and, in fact, covers those energy ranges that are not available from any other source. Thus, synchrotron radiation allows the researcher to choose any desired photon energy for the experiment, within the limits set by the monochromator. Again the dipole radiation emitted in the orbital plane of the electron storage ring is linearly polarized, with the electric field vector parallel to the plane of the orbit. Besides this, it is also possible to generate circularly polarized radiation. Such high degree of polarization is very useful in many spectroscopic and structural experiments for exploring the electronic structure and the spatial symmetry of systems under investigation. Pulsed time-structure of the radiation allows one to probe the infrared dynamics of the systems investigated.

A Typical Beamline for NEXAFS Studies

The BEAR beamline [150] at the ELETTRA synchrotron is a very good example of NEXAFS studies of soft materials bearing low-Z

Figure 3.21: *The BEAR beamline at Elettra Synchrotron, Trieste. (a) side view of preparation chamber, and (b) close view of characterization chamber.*

atoms. The apparatus is based on a bending magnet as a source. The beamline optics delivers photons from 4 eV up to about 1600 eV with selectable degree of ellipticity. The UHV end-station has a movable hemispherical electron-energy analyzer and a set of photodiodes to collect angle-resolved photoemission spectra, optical reflectivity and fluorescence yield, respectively. The photoemission/scattering geometries can be chosen with a wide flexibility due to movability of the detectors within the UHV chamber. The experimental station includes the two UHV chambers: (a) the preparation and (b) characterization chambers (Fig. 3.21), coupled to each other in UHV. The first chamber is for preparation of surfaces, interfaces and multi-layers and their in situ electronic and structural characterization. Samples can be inserted from the atmosphere and moved to the different stations of the preparation and characterization chambers (UPS and AES, LEED, evaporation/deposition, scanning tunnelling microscopy (STM)) and transferred to/from the spectroscopy chamber. The spectroscopy chamber features optical reflectivity and optical absorption, EXAFS, NEXAFS, as well as electron spectroscopies including band mapping by photoemission, high resolution core-level photoemission and photoelectron diffraction.

Chapter 4

Theoretical Background for Data Analysis

In this chapter, we have briefly discussed the theory required for quantitative analysis of our experimental data. These include height-height correlation functions which are extracted from the Atomic Force Microscopy (AFM) data, the Parratt formalism used to fit the X-Ray Reflectivity (XRR) data and obtain electron density profiles along sample depth and the theory of normal coordinates, which is used for quantitative analysis of Fourier Transform Infrared (FTIR) spectroscopy data to get information regarding bonding, coordination and conformation in molecules. we have also briefly discussed the useful relation regarding Near Edge X-Ray Absorption Fine Structure spectroscopic data analysis to obtain spatial distribution of bonds in our systems.

4.1   Analysis of Atomic Force Microscopy Data

AFM data have been quantitatively analyzed to obtain the height-height correlation functions from the topographic images. In this section, we have briefly described the relevant correlation functions that are used to explain the observed density variation in soft matter thin films.

4.1.1   Height–Height Correlation Functions

Correlation of a rough surface

All rough surfaces exhibit fluctuations along the surface normal, which are characterized by a mean-square roughness $\sigma = \overline{h(x_0, y_0)^2}^{1/2}$, $h(x_0, y_0) = h_0(x_0, y_0) - \overline{h_0(x_0, y_0)}$ where $h_0(x_0, y_0)$ is the height at any point (x_0, y_0) and the overline denotes spatial average over a planar rough surface [151, 152]. The roughness is termed "Gaussian" if $h(x_0, y_0) - h(x', y')$ is a Gaussian random variable whose distribution depends only on the relative coordinates $(x, y) = (x' - x_0, y' - y_0)$. For any isotropic Gaussian rough surface , the mean-square surface fluctuation is given by the height-difference correlation function

$$g(r) =< [h(x + x_0, y + y_0) - h(x_0, y_0)]^2 > \qquad (4.1)$$

where $r^2 = (x^2 + y^2)$, the average is taken over all pairs of points on the surface which are separated horizontally by the length r, with $<>$ denoting an ensemble average over all possible roughness configurations.

Self-affine roughness

If the surface exhibits self-affine roughness, $g(r)$ will scale as $g(r) \propto r^{2H}$, where H is called the "roughness (Hurst) exponent." The mean-square roughness of any physical self-affine surface will saturate at sufficiently large lengths in plane. It is characterized by a correlation length ξ such that,
$g(r) \propto r^{2H}$ for $r \ll \xi$,
and
$g(r) = 2\sigma^2$ for $r \gg \xi$,
where σ is the saturation roughness.
A self-affine rough surface is characterized by [151]

$$g(r) = 2\sigma^2(1 - exp(-(r/\xi)^{2H})) \qquad (4.2)$$

LB multilayers of metal-bearing fatty acids typically exhibit this type of self-affine roughness. We have used Eqn. 4.2 to characterize the in-plane morphology of cadmium stearate multilayers in Chapter 8 of our book.

Liquid-like correlation

A liquid-vapor or liquid-liquid interface possesses the property that the density-density correlations induced by capillary-wave fluctuations at the interface are long ranged (i.e., macroscopic), in contrast to density fluctuations in the bulk liquid, which are short ranged. This is a remarkable consequence of dimensionality [153] and leads to a long range height-height correlation. From the theory of capillary waves, if h(r) is the z component of the interface displacement at lateral position r relative to some arbitrary origin, the quantity $g(r) = <[h(r)h(0)]^2>$ is given to a good approximation by [132],

$$g(r) = 2\sigma_{in}^2 - BK_0(\kappa r) \qquad (4.3)$$

where $B = k_B T_0/\pi\gamma$, γ is the surface tension at temperature T_0 (8=6.0 Å^2 for ethanol at room temperature), $K_0(x)$ is the modified Bessel function (Kummer function) and σ_{in}^2 is the total intrinsic mean-square surface displacement. In particular,

$$\sigma_{in}^2 = \frac{1}{4}Bln[(q_u^2 + \kappa^2)/\kappa^2] \qquad (4.4)$$

where κ is the gravitational cutoff (lower wave vector cut-off) given by $(\Delta\rho_m g_a/\gamma)^{1/2}$, $\Delta\rho_m$ being the difference in mass density across the interface, and g_a is the acceleration due to gravity whereas q_u is the upper wave-vector cutoff for the capillary waves, given by the molecular size.

Liquid-like correlations in LB films

Height-height correlation studies using x-ray scattering on divalent metal-bearing amphiphilic fatty acid Langmuir–Blodgett multilayers showed that they have liquid-like height correlation when the observation area is large [57,65,72,73,154,155]. On the other hand, the same multilayers are shown to exhibit self-affine correlation, i.e., a non-liquid-like behavior, over small scan lengths [72, 73] as observed from AFM studies.

In our book, we have shown from AFM studies, that certain divalent metal-bearing fatty acid LB multilayers do exhibit "liquid-like" correlation at small scan lengths. In order to explain this behavior, we have used a new logarithmic function $f(r)$ of the form

$$f(r) = d_p \ ln(a_p r^2 + b_p r + c_p) \tag{4.5}$$

where a_p, b_p, c_p and d_p are parameters. Although the logarithmic dependence observed for capillary waves generated on a liquid surface [57,132] has a linear dependence on r along with some constant, our equation contains a term quadratic in r, the coefficient of which is probably related to the areal number density, and a linear term in r with a coefficient related to the wavenumber of the capillary waves, c_p and d_p being two constants involved with self-correlation.

4.2 Analysis of X-Ray Reflectivity Data

4.2.1 The Parratt Formalism

The basis of X-rays and their elastic scattering from surfaces and interfaces have been discussed in Chapter 3. Here we will discuss details of the analysis of X-Ray reflectivity at grazing incidence in order to obtain information about the materials at whose surface the reflection is taking place. XRR data of thin films or multilayers is analyzed to obtain their *Electron Density Profiles (EDPs)*, which give the variations of average electron densities of the materials along their depths. EDPs are obtained by using a recursive formula due to Parratt [156], which is based on solving Fresnel equations at multiple interfaces.

Reflectivity from multiple interfaces

We consider a film of thickness d on a substrate. A typical reflectivity profile, as is shown in Fig. 3.9, contains Kiessig fringes, i.e., interference fringes originating from the variation in electron density. The difference between successive minima of these fringes is inversely related to the thickness of the film. We consider air, film and the substrate to be denoted by media 0, 1 and 2 respectively. If the reflection coefficients at the air-film interface ($z = 0$) and

Figure 4.1: *Schematic of Parratt formalism.*

film-substrate interface ($z = d$) are denoted by r_{01} and r_{12} respectively, then the reflectance from this air-film-substrate system can be written as

$$r' = \frac{r_{01} + r_{12}}{1 + r_{01}r_{12}} \qquad (4.6)$$

The continuity condition at the substrate-film interface at $z = d$ gives rise to an extra exponential factor in the expression of r_{12} as

$$r_{12} = \frac{k_{z,1} - k_{z,2}}{k_{z,1} + k_{z,2}} \exp(-2ik_{z,1}d) = \frac{q_{z,1} - q_{z,2}}{q_{z,1} + q_{z,2}} \exp(-iq_{z,1}d) \qquad (4.7)$$

where $q_z = 2k_z = (4\pi/\lambda)sin\theta_i$, λ being the wavelength of x-ray and θ_i is the angle of incidence (with the surface) as described in Chapter 3.

The above calculation can be extended to the case of reflectivity for stratified homogeneous media having M such thin layers [156]. Starting from air/vacuum as medium 0, the topmost layer is denoted as medium 1 and so on. Thus M-th layer is the lower most layer and $(M+1)$th layer is the substrate (Fig. 4.1). Let the thickness of n-th layer be denoted by d_n. A set of equations similar to Eqn. (4.6) can be obtained for each interface and one can arrive at a recursive formula given by [50, 134–137, 157]

$$r_{n-1,n} = \frac{r_{n,n+1} + F_{n-1,n}}{1 + r_{n,n+1}F_{n-1,n}} exp(-iq_{z,n-1}d_{n-1}) \qquad (4.8)$$

where

$$F_{n-1,n} = \frac{q_{z,n-1} - q_{z,n}}{q_{z,n-1} + q_{z,n}} \qquad (4.9)$$

The reflectance is first calculated using Eqn.(4.8) between the layer closest to the substrate and the substrate, i.e., the Mth and ($M+$

$1)^{th}$ layer respectively with the assumption that $r_{M+1,M+2} = 0$, as the thickness of this layer (substrate) is taken as infinite. The calculated reflectivity is then fit to the experimental data to obtain the EDP of the multilayer structure.

The above calculation is not only used for multilayer thin films, but also widely utilized to approximate realistic electron density profile of a monolayer film. This is done by subdividing the continuous electron density profile of the film into a series of (imaginary) discrete layers each having thickness d_n and electron density ρ_n. The reflectivity of this layered system is calculated using the above mentioned recursion relation. The reflectivity profile calculated using this model is modified iteratively until it matches the experimental data. The layer thickness and electron densities are used to construct the electron density profile along film depth.

Effect of roughness

So far we have discussed the reflectivity profile for ideally smooth surfaces/interfaces. For a real system one has to include the effect of roughness in reflectance of each interface which reduces the reflectivity of the surface by scattering x-rays in non-specular direction. In most of the cases, it is assumed that deviations of the actual interface from the average value of the interface can be described by gaussian statistics. This is equivalent to convoluting the infinitely sharp profile with a Gaussian smoothing function. If roughness at the interface separating media n-1 and n is considered to be σ_n i.e., the standard deviation of the actual interface from the average value of the interface is σ_n, Eqn. (4.8) will be modified as [50],

$$r_{n-1,n}^{rough} = r_{n-1,n} \exp(-\frac{1}{2} q_{z(n-1)} q_{zn} \sigma_n^2), \qquad (4.10)$$

These reflectances can then be used to calculate the reflectivity of the entire film in the above mentioned recursive formula iteratively. The reflectivity from a surface having roughness σ is given by

$$R_{rough} = R_F \exp(-q_z^2 \sigma^2) \qquad (4.11)$$

where R_F is the Fresnel reflectivity given by Eqn. (3.32). It is clear from Eqn. (4.10) and Eqn. (4.11) that the reflectivity from rough surfaces and interfaces is smaller compared to the smooth surfaces and interfaces, and this deviation increases as q_z increases. The interfacial roughness reduce the density contrast between two layers and thus reduce the amplitude of the oscillation in the reflectivity profile whereas surface roughness reduces the total reflectivity. The x-y average EDP as a function of depth, $\rho(z)$ can be written as [137]

$$\rho(z) = \sum_1^M \Delta\rho_n f_{er}(z_n, \sigma_n) \tag{4.12}$$

where $\Delta\rho_n$ is the change in electron density at n^{th} smooth interface located at a position z_n and f_{er} is an error function given by

$$f_{er}(z_n, \sigma_n) = f_{er}(z - z_n, \sigma_n) = \frac{1}{\sqrt{2\pi}\sigma_n} \int_{-\infty}^{z-z_n} \exp(-t^2/2\sigma_n^2)dt \tag{4.13}$$

where σ_n is the roughness of the n-th interface which is a parameter for the estimation of interfacial width. It can be noted that these error functions of Eqn. (4.12) come as Debye-Waller like factors in the expression of reflectance of each interface in reciprocal space as given in Eqn. (4.10).

Resolution Function

The calculated reflectivity profile cannot be compared directly with the experimental reflectivity data because one has to take into account the finite resolution of the experimental setup in the calculation [50, 135]. The resolution in reflectivity measurements depends on the wavelength dispersion $(\Delta\lambda/\lambda)$, the angular divergence of incoming beam $(\Delta\theta_i)$ and the acceptance of outgoing radiation $(\Delta\theta_f)$. Before comparing with experimental data, the calculated profile has to be convoluted with a resolution function which is generally modeled as a Gaussian function of the type

$$\Re(q_z) = \frac{1}{\sqrt{2\pi}\sigma_R} \exp\left(-\frac{(q_z - q_z')^2}{2\sigma_R^2}\right) \tag{4.14}$$

The standard deviation σ_R is related to the full-width at half-maximum (FWHM) Δq_z of a Gaussian function as $\sigma_R = \Delta q_z/(2\sqrt{2\log 2})$ [158]. This Δq_z depends on wavelength dispersion as well as on the geometry of experimental setup. In the usual geometry of the experimental setup, the aperture of the detector slits perpendicular to the scattering plane is kept wide open so that the scattered intensity in the q_y direction is integrated out in the XRR experiment. Let us recall the expressions for q_z and q_x given in Chapter 3. Assuming x-z plane as the scattering plane, the total differential of Eqn. (3.13) and Eqn. (3.15) is given by,

$$dq_z = k_0(\cos\theta_i\, d\theta_i + \cos\theta_f\, d\theta_f) + dk_0(\sin\theta_i + \sin\theta_f), \quad (4.15)$$

$$dq_x = k_0(\sin\theta_i\, d\theta_i - \sin\theta_f\, d\theta_f) - dk_0(\cos\theta_i - \cos\theta_i) \quad (4.16)$$

With the assumption that $d\theta_i$ and $d\theta_f$ are randomly distributed the resolution widths in q_x and q_z are given by,

$$\Delta q_z^2 = k_0^2(\cos^2\theta_i\, \Delta\theta_i^2 + \cos^2\theta_f\, \Delta\theta_f^2) + \Delta k_0^2(\sin\theta_i + \sin\theta_f)^2, \quad (4.17)$$
$$\Delta q_x^2 = k_0^2(\sin^2\theta_i\, \Delta\theta_i^2 + \sin^2\theta_f\, \Delta\theta_f^2) + \Delta k_0^2(\cos\theta_i - \cos\theta_f)^2 \quad (4.18)$$

In a standard geometry the distance between the x-ray source and sample was kept equal to that between the sample and the detector. In this geometry $\Delta\theta_i = \Delta\theta_f = \Delta\theta$ (say), which can be obtained from the FWHM of the direct beam profile. Then in specular condition above two equations can be rewritten as,

$$\Delta q_z = E\left[\frac{1}{2}\cos^2\theta\, \Delta\theta^2 + \left(\frac{\Delta E}{E}\right)^2 \sin^2\theta\right]^{1/2} \quad (4.19)$$

$$\Delta q_x = E\left[\frac{1}{2}\sin^2\theta\, \Delta\theta^2 + \left(\frac{\Delta E}{E}\right)^2 \cos^2\theta\right]^{1/2} \quad (4.20)$$

For the monochromatic X-Ray source, we do not have to consider the effect of wavelength dispersion and hence the second term in the above equations will vanish and only the first terms in expression of Δq_z and Δq_x will contribute to the resolution function.

4.3 Analysis of FTIR Data

4.3.1 Normal Coordinate Analysis

Molecular Vibrations

A molecular vibration occurs when atoms in a molecule are in periodic motion while the molecule as a whole has constant translational and rotational motion. The frequency of the periodic motion is known as a vibration frequency, and the typical frequencies of molecular vibrations range from less than 10^{12} to approximately 10^{14} Hz. In general, a molecule with N atoms has 3N-6 normal modes of vibration but linear molecules have only 3N-5 normal modes of vibration as rotation about its molecular axis cannot be observed [159]. A diatomic molecule has one normal mode of vibration. The normal modes of vibration of polyatomic molecules are independent of each other but each normal mode will involve simultaneous vibrations of different parts of the molecule such as different chemical bonds.

A molecular vibration is excited when the molecule absorbs a quantum of energy, E, corresponding to the frequency of vibration, ν, according to the relation E = hν, where h is Planck's constant. A fundamental vibration is excited when one such quantum of energy is absorbed by the molecule in its ground state. When two quanta are absorbed the first overtone is excited, and so on to higher overtones. To a first approximation, the motion in a normal vibration can be described as a kind of simple harmonic motion. In this approximation, the vibrational energy is a quadratic function (parabola) with respect to the atomic displacements and the first overtone has twice the frequency of the fundamental. In reality, vibrations are anharmonic and the first overtone has a frequency that is slightly lower than twice that of the fundamental. Excitation of the higher overtones involves progressively less and less additional energy and eventually leads to dissociation of the molecule, as the potential energy of the molecule is more like a Morse potential.

The vibrational states of a molecule can be probed in a variety of ways. The most direct way is through infrared spectroscopy, as vibrational transitions typically require an amount of energy that corresponds to the infrared region of the spectrum. Vibrational excitation can occur in conjunction with electronic excitation (vibronic transition), giving vibrational fine structure to electronic transitions, particularly with molecules in the gas state. Simultaneous excitation of a vibration and rotations gives rise to vibration-rotation spectra.

Vibrational Coordinates

Internal Coordinates: As the name suggests, these coordinates (q) describe the internal configuration of the molecule without regard for its position as a whole in space. These include (a) *stretching*: a change in the length of a bond, such as C-H or C-C; (b) *bending*: a change in the angle between two bonds, such as the H-C-H angle in a methylene group; (c) *rocking*: a change in angle between a group of atoms, such as a methylene group and the rest of an ethylene molecule; (d) *wagging*: a change in angle between the plane of a group of atoms, such as a methylene group and a plane through the rest of the molecule; (e) *twisting*: a change in the angle between the planes of two groups of atoms, such as a change in the angle between the two methylene groups and (f) *out-of-plane*: a change in the angle between any one of the C-H bonds and the plane defined by the remaining atoms of an ethylene molecule.

Normal coordinates: The coordinate of a normal vibration is a combination of changes in the positions of atoms in the molecule [90, 160, 161]. When the vibration is excited the coordinate changes sinusoidally with a frequency ν, the frequency of the vibration. The normal coordinates, denoted as Q, refer to the positions of atoms away from their equilibrium positions, with respect to a normal mode of vibration. Each normal mode is assigned a single normal coordinate, and so the normal coordinate refers to the "progress" along that normal mode at any given time. Formally, normal modes are determined by solving a secular determinant, and then the normal coordinates (over the normal modes) can be expressed as a summation over the cartesian coordinates (over the atom positions). The advantage of working in normal modes is that they diagonalize the matrix governing the molecular vibrations, so each normal mode is an *independent molecular vibration*, associated with its own spectrum of quantum mechanical states. If the molecule possesses symmetries, it will belong to a point group, and the normal modes will "transform as" an irreducible representation under that group. The normal modes can then be qualitatively determined by applying group theory and projecting the irreducible representation onto the Cartesian coordinates.

For example, when this treatment is applied to CO_2, it is found that the C=O stretches are not independent, but rather there is a O=C=O symmetric stretch and an O=C=O asymmetric stretch. In case of symmetric stretching, the two C-O bond lengths change by the same amount and the carbon atom is stationary, such that the normal mode is the sum of the two C-O stretching coordinates $Q = q_1 + q_2$. On the other hand, in asymmetric stretching, one C-O bond length increases while the other decreases and the difference of the two C-O stretching coordinates i.e., $Q = q_1 - q_2$ gives the normal mode.

When two or more normal coordinates belong to the same irreducible representation of the molecular point group (i.e., have the same symmetry) there is "mixing" and the coefficients of the combination cannot be determined a priori. These coefficients are found by performing a full Normal Coordinate Analysis (NCA) by means of the Wilson GF method [160].

Normal Coordinate Analysis: GF Method

The GF method, is a classical mechanical method introduced by E. Bright Wilson to obtain certain internal coordinates for a vibrating semi-rigid molecule, the so-called normal coordinates Q_k. Normal coordinates decouple the classical vibrational motions of the molecule and thus give an easy route to obtaining vibrational amplitudes of the atoms as a function of time. In Wilson's GF method it is assumed that the molecular kinetic energy consists only of harmonic vibrations of the atoms, i.e., overall rotational and translational energy is ignored. Normal coordinates appear also in a quantum mechanical description of the vibrational motions of the molecule and the Coriolis coupling between rotations and vibrations. It follows from application of the Eckart conditions that the matrix $\mathbf{G}^{-1}$ gives the kinetic energy in terms of arbitrary linear internal coordinates, while $\boldsymbol{F}$ represents the (harmonic) potential energy in terms of these coordinates. The GF method gives the linear transformation from general internal coordinates to the special set of normal coordinates.

A non-linear molecule consisting of N atoms has 3N-6 internal degrees of freedom, because positioning a molecule in three-dimensional space requires three degrees of freedom and the description of its orientation in space requires another three degrees of freedom. These degrees of freedom must be subtracted from the 3N degrees of freedom of a system of N particles. The atoms in a molecule are bound by a potential energy surface (PES) (or a force field) which is a function of 3N-6 coordinates. The internal degrees of freedom $q_1, ..., q_{3N-6}$ describing the PES in an optimum way are often non-linear; they are for instance valence coordinates, such as bending and torsion angles and bond stretches. It is possible to write the quantum mechanical kinetic energy operator for such curvilinear coordinates, but it is hard to formulate a general theory applicable to any molecule. Wilson linearized the internal coordinates by assuming small displacements.

Even then, NCA becomes tedious for a complex molecular system. In such cases, the empirical approach is employed in which, various sub-parts of the molecule are treated separately. NCA of these functional groups are carried out, taking into account their molecular environment. This is possible due to the fact that various functional groups present in a molecule give rise to absorption bands in the same spectral region (commonly called *Group Frequencies*).

NCA: Empirical Approach

In this approach, individual functional groups of the molecule are treated as isolated systems (typically diatomic or triatomic systems), and their vibrational frequencies calculated by NCA. For a diatomic molecule comprising atoms **A** and **B**, having masses M_A and M_B, the frequency of vibration (expressed in terms of its wavenumber $\bar{\nu}$) is given by [162],

$$\bar{\nu} = \frac{1}{2\pi c}\sqrt{\frac{k_s}{\mu_m}} \tag{4.21}$$

where $\mu_m^{-1} = (M_A^{-1} + M_B^{-1})$, and k_s is the spring constant of the bond between the molecules.

For a non-linear symmetric triatomic molecule **A-B-A** such that atom **B** is bonded to two **A** atoms, there are three normal modes of vibration; one antisymmetric and two symmetric. The antisymmetric mode is a stretching modes having frequency $\bar{\nu}_a$ given by,

$$\bar{\nu}_a = \frac{1}{2\pi c}\sqrt{\frac{k_s}{\mu_m}\left(1 + \frac{2M_A}{M_B}sin^2\alpha\right)} \tag{4.22}$$

The frequencies $\bar{\nu}_{si}, i = 1, 2$ of the other two symmetric modes (i.e., symmetric stretch and bend) can be obtained from the roots $\omega_{s1,s2}$ of the equation

$$\omega_{s1,s2}^2 = \frac{1}{2}\left[t_1 + t_2 \pm (t_1 + t_2)^2 - \left(\frac{8\mu_m}{M_A^2 M_B}k_s k'\right)^{1/2}\right] \tag{4.23}$$

such that

$$t_1 = \frac{k_s}{M_A}\left(1 + \frac{2M_A}{M_B}cos^2\alpha\right) \tag{4.24}$$

and

$$t_2 = \frac{2k'}{M_A}\left(1 + \frac{2M_A}{M_B}sin^2\alpha\right) \tag{4.25}$$

where $\bar{\nu} = \omega/(2\pi c)$, k' being the spring constant corresponding to bond bending.

The Coupling Effect

The origin of group frequency is that a particular functional group has frequencies in the same spectral region and is independent of

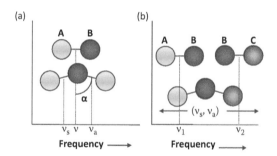

Figure 4.2: *Illustration of coupling effect between (a) same groups and (b) different groups.*

the molecular environment. However in reality, the presence of other molecules or functional groups do have appreciable effect on the frequency of vibration, *Coupling effect* [163] being the most important (Fig. 4.2).

Coupling of same groups: If in a complex system comprising a diatomic molecule **A-B** as a functional group having a frequency of vibration $\bar{\nu}$, a second atom **A** gets attached to the molecule **B**, the frequency $\bar{\nu}$ is no longer present in the spectra of the molecule. Instead, two frequencies $\bar{\nu}_a$ and $\bar{\nu}_s$ are found to exist on either side of $\bar{\nu}$, $\bar{\nu}_a$ being on the higher side of the frequency axis. This is because of the coupling between the two identical diatomic systems **A-B** to yield an isolated triatomic system **A-B-A**, such that the new frequencies $\bar{\nu}_a$ and $\bar{\nu}_s$ correspond to the asymmetric and symmetric stretch modes.

Coupling of different groups: If a functional group **A-B-C** is present in a complex molecule, the two systems **A-B** and **B-C** having frequencies $\bar{\nu}_1$ and $\bar{\nu}_2$, can only be considered to be isolated diatomics if their frequencies of vibration are widely separated from each other. However, if $\bar{\nu}_1$ and $\bar{\nu}_2$ lie close to each other on the frequency axis, the system is coupled and is to be treated as an asymmetric triatomic molecule instead.

Seibert Relations

There are several empirical relations to calculate the force constant of bond stretch, of which the one due to Seibert is most useful.

Here, the force constant k_{A-B} of a single bond **A-B** is given by [164],

$$k_{A-B} = \frac{7.20 Z_A Z_B}{(n_A n_B)^3} \tag{4.26}$$

where $Z_A (Z_B)$ is atomic number, $n_A (n_B)$ is the principle quantum number of valence electrons of atom A (B) and k_{A-B} is in N/cm. For a multiple bond of bond order N, the force constant k_N is then given by,

$$k_N = k_{A-B} N r_1 / r_N \tag{4.27}$$

where $r_1 (r_N)$ is the order of single (multiple) bond. However it should be mentioned that these relations are only approximate and differ from experimental values by about 0.4 N/cm. We have used these relations to theoretically calculate the force constants of the C-O and C=O bonds in Chapter 5 of our book.

4.4   Analysis of NEXAFS Data

During NEXAFS measurements, the background signal was recorded by measuring the dark signal of the photodiode. The normalization of the measurements took into account the monitor current (I_O) from a 100V biased W-mesh, for possible fluctuations of the photon beam. The experimental transmittance T was then obtained by the relation,

$$T = \left(\frac{I_D - I_{D,B}}{I_0 - I_{0,B}} \right) / \left(\frac{I_D^D - I_{D,B}^D}{I_0^D - I_{0,B}^D} \right) \tag{4.28}$$

where I_D and I_D^D represent the diode currents when the beam passes through the sample and the direct beam respectively. $I_{D,B}$ and $I_{D,B}^D$ represent the background diode currents with sample and without sample respectively. Similarly I_O and I_D^O, $I_{O,B}$ and $I_{O,B}^D$ represent the mesh currents in same convention. All NEXAFS data were plotted as transmittance versus photon energy curves.

Chapter 5

Headgroup Coordination and Film Structure

One very important question in Langmuir film research is whether it is possible to tune film morphology by manipulating the headgroup structure via subphase metal ions. To answer this we have chosen transition elements as the metals in metal–carboxylate LB films. It is known that divalent metal ions (those which fall in the same group as the transition elements of our choice) interact with the carboxylate headgroup and form coordination complexes [161]. Then the headgroup structure is probably decided by the type of metal–carboxylate coordination present and consequently a change in the latter would possibly cause a change in the former.

To start with, we have studied the formation of coordination complexes at the air–water interface and their selection by interfacial forces during transfer onto substrate. We have chosen metal stearates of two divalent transition elements, viz. Cadmuim stearate (CdSt) and cobalt stearate (CoSt) and studied their structural evolution as regards metal ion-headgroup coordination starting from the air–water interface to their stacking as multilayers on hydrophillic substrates. Results show that selection of coordination complexes during monolayer transfer is dependent on the transition metal ion present in subphase. Next, we have studied the morphology and structure of these transferred LB multilayers and showed that the film morphology is dependent on the headgroup coordination of the multilayer. These findings are discussed in this chapter.

5.1   Coordination Selection at Air–Water Interface

In this section, we have presented results of FTIR spectroscopy to elucidate metal–carboxylate coordinations of (a) the Langmuir monolayers at air–water interface and (b) transferred monolayers on the substrate, of CdSt and CoSt. Results show that on water surface, there is co-existence of all possible metal–carboxylate coordinations, viz. unidentate, bidentate bridged and bidentate chelate, [90, 161] along with some trace of undissociated acid. However, after transfer onto a solid substrate, not all coordinations are found to exist in the transferred monolayers; rather coordination selections occur depending upon the metal ion type. Specifically, coordination of the headgroup in Co-bearing monolayer is found to be a mixture of unidentate and bidentate bridged with the latter showing a majority while in Cd-bearing monolayer, a mixture of unidentate and bidentate bridged coordinations are observed; with the unidentate coordination showing a majority.

5.1.1   Carboxylate Headgroup of Langmuir Monolayers

Preparation and IRAS measurement: The stearic acid Langmuir Monolayers (LMs) were directly prepared in the mini Langmuir trough attached to the reflection mode accessary of the FTIR spectrometer, for Infrared Reflection Absorption Spectroscopy (IRAS) measurements as discussed in Chapter 3. For this, at first, solutions of chlorides of respective metal ions were prepared in pure water (resistivity 18.2 MΩ-cm) with pH adjusted to 6-6.5 by sodium bicarbonate. The solutions were then poured in the trough to form the subphase. Next, stearic acid ($C_{17}H_{35}COOH$) was added from a chloroform solution, to form the monolayer. The spreading volume was calculated such that when compressed till a particular barrier position (i.e., for a known trough area), the stearic acid monolayer is in its most condensed phase with area per molecule ~ 21 Å^2, as the mini trough did not have a surface pressure measuring set-up, and the same could only be obtained from the π_s-A isotherms of stearic acid with the respective metal ions in subphase taken under identical conditions separately in a Langmuir trough. The measured π_s-values were 28.5 mN/m and 31.3 mN/m for StA with Cd and Co ions in subphase, respectively. For brevity, StA monolayer in presence of Cd and Co ions are called Cd-StA and Co-StA respectively. FTIR spectra of both M-StA (M = Cd or Co) were taken under specular condition with an incident angle of 70° from normal, at a resolution of 4 cm^{-1}, in order to obtain maximum count. A liquid N_2-cooled Mercury Cadmium Telluride (MCT) detector was used to maintain a high Signal to noise (S/N) ratio.

FTIR spectra of StA, Cd-StA and Co-StA (after performing Krammers–Kronnig correction [146]) were analyzed to obtain information about the different metal–carboxylate coordinations occurring at the air/water interface. For that, the typical carboxylate (COO) stretching frequency region (~ 1300 cm^{-1} to 1900 cm^{-1}) of the individual spectra were fit with gaussian function. Although in this region of the spectra, a series of equally intense sharp peaks due to methylene deformation modes (scissor, rock, wag, twist) show up, yet they do not mask the broad carboxylate stretch modes. Thus, in fitting the spectra, contributions from these deformation modes have been neglected, and only the COO modes have been assigned.

The aforementioned portion of the spectra for StA (Fig. 5.1), Cd-StA (Fig. 5.2(a)) and Co-StA (Fig. 5.2(b)) are shown and the assigned COO modes are provided in Table 5.1. Spectra of both M-StA are more or less same with seven prominent peaks in each. These are compared with the five fit peaks of StA on water without any metal ion in the subphase.

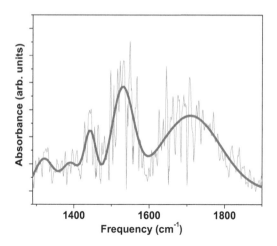

Figure 5.1: *Portion of FTIR spectra (thin solid line) of stearic acid monolayer on water showing the carboxylate stretch modes. Bold solid line is the composite fit.*

A. IR spectra of stearic acid monolayer

The StA spectra (Fig. 5.1) shows five prominent peaks, two strong ones at 1529 cm^{-1} and 1709 cm^{-1} corresponding to the carboxylate asymmetric stretch modes and two peaks of comparatively lesser intensity at 1320 cm^{-1} and 1391 cm^{-1} corresponding to the carboxylate symmetric stretch modes. The peak pairs 1709 cm^{-1} and 1320 cm^{-1} correspond to the asymmetric and symmetric stretch modes of undissociated acid, whereas the peak pairs 1529 cm^{-1} and 1391 cm^{-1} correspond to same for dissociated acid, i.e., for the carboxylate ion at air/water interface. Difference in symmetric (ν_s) and asymmetric (ν_a) COO stretching frequencies (Δ) for both undissociated and dissociated acid was estimated. For the former, (Δ) = 389 cm^{-1} whereas for the latter (Δ) = 138 cm^{-1}. A fifth peak of medium intensity, observed at 1444 cm^{-1} is attributed to the methylene scissor mode.

B. IR spectra of metal stearate monolayers

As seen from Table 5.1, both Cd-StA (Fig. 5.2(a)) and Co-StA (Fig. 5.2(b)) spectra have been fit with seven prominent peaks.

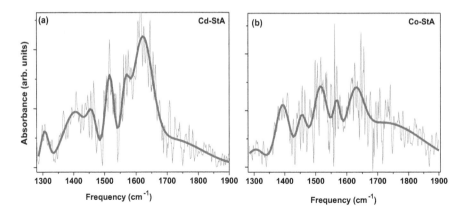

Figure 5.2: *Portion of FTIR spectra (thin solid line) showing carboxylate stretch modes of stearic acid monolayer on water containing (a) Cd and (b) Co ions in subphase. Bold solid line is the composite fit.*

The peak pairs corresponding to undissociated acid is present in both cases, although in considerably lesser amount compared to that of StA, with the symmetric stretch mode (1320 cm^{-1}) downshifted by about 15 cm^{-1}. The methylene scissor mode (1444 cm^{-1} in StA), also present in Cd-StA and Co-StA spectra at 1461 cm^{-1} and 1455 cm^{-1} respectively, show a slight frequency upshift (17 cm^{-1} for CdSt and 11 cm^{-1} for CoSt) suggesting that tails are fixed more rigidly to the headgroup but as they have more or less the same intensity as that of StA, there is probably very small interaction of tails due to change in headgroup structure, owing to presence of metal ions in subphase.

However, the symmetric and asymmetric stretch modes of dissociated acid, i.e., the carboxylate ion are different for the two metals. In order to better understand the dissociated acid peaks, the same are plotted together for both metal ions and for StA (Fig. 5.3). Both symmetric and asymmetric stretch modes in metal stearates are thus seen to be shifted and considerably enhanced in intensity. To elaborate, in stearic acid, the COO symmetric stretch is observed at 1391 cm^{-1}. The same is observed at 1405 cm^{-1} and 1393 cm^{-1} for Cd-StA and Co-StA respectively. Similarly, the COO asymmetric stretch, observed at 1529 cm^{-1} in StA, is not

Table 5.1: IRAS data of stearic acid monolayers on water

| Carboxylate stretch modes (ν cm^{-1}) of | | | Peak |
StA	Cd-StA	Co-StA	Assignment
1320	1306	1305	ν(C-O)
1391	1405	1393	ν_s(COO)
1444	1461	1455	δ(CH$_2$) scissor
-	1516	1515	ν_{a1}(COO)
1528	1566	1568	ν_{a2}(COO)
-	1621	1628	ν_{a3}(COO)
1709	1705	1718	ν(C=O)

a: asymmetric; s: symmetric; δ: deformation

found in the other MSt spectra. Instead three peaks, all assigned to COO asymmetric stretch modes (viz. ν_{a1}, ν_{a2} and ν_{a3}) are observed in both MSt spectra. For example, in Cd-StA, $\nu_{a1} = 1516$ cm^{-1}, $\nu_{a2} = 1566$ cm^{-1} and $\nu_{a3} = 1621$ cm^{-1}. Similarly $\nu_{a2} = 1569$ cm^{-1}, 1568 cm^{-1} and 1554 cm^{-1} for Co-StA.

The differences in asymmetric and symmetric stretching frequencies (Δ) were calculated for StA and both MSt, giving one value ($\Delta = 138$) for the former and three sets of values (Δ_1, Δ_2 and Δ_3) for the latter. All values of Δ are given in Table 5.2. As seen from the table, Δ_1 corresponds to bidentate chelate coordination whereas Δ_3 corresponds to unidentate coordination .

The assignment of Δ_2 is a little ambiguous, since it can indicate both a bidentate bridged coordination or an ionic bonding. In case of a bidentate bridged coordination, the Δ value is usually close to that of the dissociated carboxylate ion, as observed in bulk samples. In our case, $\Delta = 138$ whereas $\Delta_2 \sim 170$, the latter typically corresponding to ionic bonding [161]. It appears then (Fig. 5.3) that the intense peak at 1529 cm^{-1}, observed in StA, is rather split into two less intense peaks on either side of it (for example at 1516 cm^{-1} and 1566 cm^{-1} in Cd-StA). Finally, as the individual Δ values vary from sample to sample they are very unlikely to come from ionic bonding and hence, Δ_2 is assigned to bidentate bridged coordination rather than ionic bonding. Thus, the final observation is that at the air/water interface all types of coordinations are allowed for stearic acid with Cd and Co ions in subphase.

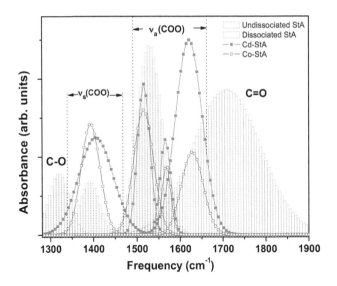

Figure 5.3: *The COO stretch modes (peak fits) of Cd-StA and Co-StA along with that of stearic acid on water.*

A preliminary calculation involving the area under the peaks for symmetric and asymmetric COO stretch modes was carried out to estimate the relative ratios of the different types of coordinations present. For this, the ratio C_i (i=1,2,3) given by

$$C_i = a_i/a_0 \tag{5.1}$$

was computed, where a_i (a_0) is the area under the peaks of asymmetric (symmetric) stretch mode of the dissociated carboxylate ion.

The C_i-values (Table 5.3) show that in Cd-StA, unidentate coordinated carboxylate groups form the majority, whereas in Co-StA, most COO groups coordinate with the Co ions via bidentate chelate coordination.

Table 5.2: Δ-values for M-StA

Δ_{ai}	Cd-StA	Co-StA
Δ_{a1}	111	122
Δ_{a2}	161	175
Δ_{a3}	216	325

Table 5.3: C_i-values for MSt

C_i	Cd-StA	Co-StA	CdSt	CoSt
C_1	0.55	1.16	-	-
C_2	0.31	0.30	0.39	0.31
C_3	1.56	0.86	0.81	0.12

5.2 Carboxylate Headgroup of Transferred Monolayers

The LB transfer process is continued beyond deposition of a monolayer on the hydrophillic substrate to yield LB multilayers of CdSt and CoSt. The metal–carboxylate bonding of these multilayers have been investigated by FTIR and NEXAFS spectroscopic techniques, and results suggest no further change in coordination from the transferred LB monolayer except for the fact that the minority coordinations show negligible contribution in both CdSt and CoSt films. In order to study the dependence of film morphology on coordination, AFM and XRR studies of the two films were carried out. Results show that the morphology is drastically changed by the change in metal ion headgroup coordination, and that Co-carboxylate coordination yields defect-free multilayers.

5.2.1 Deposition and IR Measurement

Using the LB trough, one monolayer (ML) films of cadmium stearate (CdSt) and cobalt stearate (CoSt) were deposited onto hydrophilic silicon (100) substrate by LB technique described in Chapter 2, from their respective Langmuir monolayers, the latter being formed under the same conditions as before. All substrates used for film deposition were hydrophilized by keeping them in a solution of ammonium hydroxide, hydrogen peroxide and pure water (H_2O: NH_4OH: H_2O_2 2:1:1 by volume) for 10-15 minutes at 100°C. The two 1ML films were deposited at a surface pressure of 30 mN/m (i.e., area/mol $\sim$ 21 Å^2) at 20°C at a dipping speed of 3mm/min by an upstroke of the substrate from water to air through the interface. FTIR spectra of the two films were taken in transmission mode at a resolution of 4.0 cm^{-1} using the MCT detector.

5.2.2 IR Spectra of Transferred Metal Stearate Monolayers

The spectra of 1ML CdSt (Fig. 5.4(a)) and CoSt (Fig. 5.4(b)) films have been fit with Gaussian functions and the fitted peaks are given in Table 5.4. Here again, only the COO peaks (both dissociated and undissociated) have been assigned, by comparing the spectra with that of bulk stearic acid. For clarity, the fitted spectra of CdSt and CoSt with the same for bulk StA have been plotted separately (Fig. 5.5).

As seen from the figure, both MSt spectra show presence of undissociated acid in negligible amount compared to bulk StA. The C-O stretch modes are at 1301 cm^{-1} and 1298 cm^{-1} for CdSt and CoSt respectively. Similarly, all peaks beyond 1576 cm^{-1} in both metal stearates are assigned to C=O stretch modes since they fall within the region of C=O mode in bulk StA and hence are not individually assigned. The methylene scissor mode, present at 1454 cm^{-1} (1453 cm^{-1}) in CdSt (CoSt) shows slight downshift compared to Cd-StA while this mode in CoSt remained unchanged compared to that in Co-StA. The COO symmetric stretch is observed at 1385 cm^{-1} (1382 cm^{-1}) in CdSt (CoSt). The peaks at

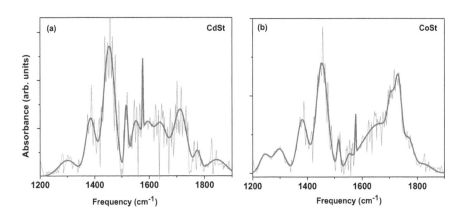

Figure 5.4: *Portion of FTIR spectra (thin solid line) showing carboxylate stretch modes of (a) CdSt and (b) CoSt monolayers on hydrophilic silicon substrate. Bold solid line is the composite fit.*

Table 5.4: FTIR data of MSt monolayers on Si

COO stretches (ν cm^{-1}) of		Assignment
CdSt	CoSt	
-	1244	ν(C-O)
1301	1298	ν(C-O)
1385	1382	ν_s(COO)
1454	1453	δ(CH$_2$) scissor
-	-	ν_{a1}(COO)
1516	1515	ν_{a2}(COO)
1548	1552	ν_{a3}(COO)
1576	1576	
1591	1666	
1640	1704	ν(C=O)
1712	1731	
1775	1770	
1844	1830	

a: asymmetric; s: symmetric; δ: deformation

1516 cm^{-1} and 1548 cm^{-1} (1515 cm^{-1} and 1552 cm^{-1}) are assigned to the COO asymmetric stretch modes in CdSt (CoSt). Thus it is found that in both CdSt and CoSt, there are two sets of values for Δ viz. Δ_1 and Δ_2, corresponding to bidentate and unidentate coordinations, in contrast to the three sets of Δ values obtained for the M-StA system, indicating that one of the bidentate coordinations (chelate) is not transferred onto the substrate. The assigned values are given in Table 5.5.

It must be mentioned here that the assignments of Δ values for the transferred films have been made by comparing the values for similar systems both from literature including our own results. However, these assignments do not strictly match those of Cd-StA and Co-StA. This is due to the fact that although coordination type is same, strength of metal–carboxylate coordination is likely to vary at air/water interface and at air/substrate interface. It is for this reason that the specific type of bidentate coordination formed on the substrate cannot be determined. Nevertheless, the ratio C_i was calculated for both CdSt and CoSt monolayers. Results show that in CdSt, unidentate coordination and in CoSt, bidentate coordination is still the majority (Table 5.3). Thus our results show that at air/substrate interface not all coordinations

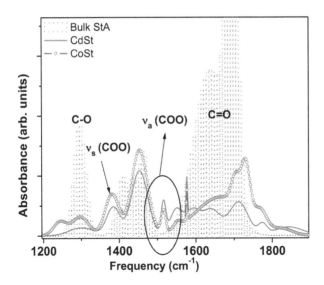

Figure 5.5: *The COO stretch modes (peak fits) of CdSt and CoSt monolayers along with that of bulk stearic acid on silicon substrate.*

Table 5.5: Δ-values for MSt

Δ_{ai}	CdSt	CoSt
Δ_{a1}	-	-
Δ_{a2}	131	133
Δ_{a3}	163	170

exist. Rather specific coordinations are selected depending upon the metal ion type.

Again, the ratio C_i was calculated for the MSt monolayers. Results show that in CdSt, unidentate coordination is still preferred whereas in CoSt, bidentate coordination is the majority. Comparing with the ratios for M-StA monolayers, it was found that during transfer, CdSt coordination preference remains practically unaltered, whereas those CoSt changes significantly. Specifically, for CoSt the coordination changed from chelate to bidentate bridged. our results thus confirm that carboxylate coordination is altered during the transfer process and is strongly related to the metal ion type. Since all coordinations were allowed at the water surface, we tentatively suggest that this change in the majority coordination on LB transfer is through a selection process, rather than active change in bonding. Dynamical studies during film transfer may clarify this further. Our preliminary results thus suggest strong and metal-specific supramolecular bonds mediated by the hydrophillic substrate surface, as a selection mechanism for the metal–headgroup coordination that is transferred. A recent simulation has highlighted the role of short-range interactions at hydrophilic substrate surface in stabilizing polyelectrolytes through polarization of water molecules hydrogen bonded to the polymer [165]. A similar mechanism may be at work here, selecting the headgroup coordination through the selection of specifically oriented water molecules attached to the headgroup during transfer.

To summarize, the formation of coordination complex by metal salts of fatty acids at air–water interface is driven by the type of metal ion present in subphase. Whereas all types of coordinations prevail at the air–water interface, the LB transfer process causes selection of specific coordination which again shows a strong dependence on metal ion type. More specifically, it is observed that the carboxylate headgroup prefers to coordinate with the Cd ions via unidentate coordination and with Co ions via bidentate coordination, both on the air–water interface and the air-substrate interface. The next step is to explore the metal carboxylate coordinations on subsequent transfer, i.e., for multilayer LB deposition. Also, a study of the multilayer morphology for both the metal stearates is essential to obtain information regarding dependence of the same on headgroup coordination. These results are discussed in the next section.

5.3   Coordination and Multilayer Morphology

5.3.1   Multilayer Deposition and Characterization

We have deposited 9 MLs of CoSt and CdSt on hydrophilic Si (100) and fused quartz substrates where the latter were used for transmission-FTIR measurements. Respective metal ions were introduced by adding solutions of their chloride salts in the Langmuir trough containing pure water at pH $\sim$ 6.0, adjusted by sodium bicarbonate. stearic acid monolayers were spread from a chloroform solution. The films were deposited at monolayer pressure of 30mN/m at 19°C with 3 mm/min dipping speed. Drying time was 10 min and 5 min after first and subsequent strokes, respectively. Films were checked for reproducibility. It should be mentioned here that although both CdSt and CoSt LB multilayers could be drawn with ease, CdSt film coverages were found to vary randomly with every deposition while CoSt film coverage remained more or less constant.

FTIR and NEXAFS was used to characterize the headgroup structure of the multilayers. FTIR spectroscopy was carried out in transmission and ATR modes. NEXAFS spectroscopy [147] at oxygen K-edge (530 eV) was carried out at normal incidence. The spectra were recorded by total electron yield (TEY). They were normalized to an I_0 spectrum previously measured for clean copper to account for oxygen in mesh. Monochromator slits were set to provide an energy resolution of 0.1 eV. The morphology and structure of multilayers are characterized by AFM and XRR. Surface morphology of films was studied by contact mode AFM. Scans were performed in constant force mode over several regions of film with a low force constant (1.3 mN) was used to minimize damage. XRR [137] was done using X-Ray wavelength $\lambda = 1.54$ Å (Cu K$_\alpha$ line).

5.3.2 Carboxylate Headgroup of Metal Stearate Multilayers

FTIR study

FTIR-ATR spectroscopy was carried out to explore primarily the metal-ion headgroup coordination and some aspects of molecular structure of these films (Fig. 5.6(b)). Assigned peaks of these and bulk stearic acid (Fig. 5.6 (a)) are tabulated below (Table 5.6) [166, 167]. Absence of OH stretch around 3400 cm^{-1} and reduction

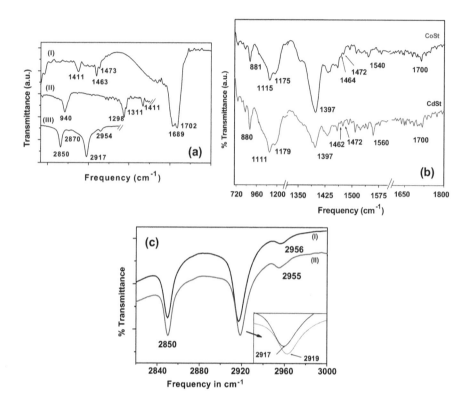

Figure 5.6: *(a) FTIR spectra of bulk stearic acid. Portions (II) and (III) of the spectra have been arbitrarily shifted for clarity. (b) FTIR spectra of (I): CoSt and (II): CdSt LB films. Y-axis for the selected frequency regions are chosen arbitrarily for better clarity. (c) FTIR spectra of (I) CoSt and (II) CdSt in transmission mode.*

in C=O stretch near 1700 cm^{-1} in LB film spectra confirm conversion of free stearic acid to corresponding metal stearate [166]. To find metal-ion carboxylate coordination, difference in symmetric (ν_s) and asymmetric (ν_a) COO stretching frequencies (Δ) are noted [161]. For CoSt ($\Delta = 143$) the coordination is bidentate bridging [168], corresponding to one carboxylate group per metal ion. However CdSt has unidentate coordination ($\Delta = 163$) with two carboxylate groups per metal ion. LB films of fatty acids are known to have orthorhombic packing with all-trans conformation of acyl chains [5]. Presence of two peaks near 1462 cm^{-1} and 1474

Table 5.6: Peak assignments of FTIR spectra of the LB films and bulk stearic acid

Observed vibrational frequencies (cm^{-1}) for			
stearic acid (Bulk)	CdSt	CoSt	Assigned to
(3750-3000)	-	-	$\nu(O-H)$
2954 (W)	2955 (W)	2956 (W)	$\nu_a(CH_3)$
2917 (S)	2919 (S)	2917 (S)	$\nu_a(CH_2)$
2870 (W)	(VW)	(VW)	$\nu_s(CH_3)$
2850 (S)	2850 (S)	2850 (S)	$\nu_s(CH_2)$
1702 (VS)	-	-	$\nu(C=O)$; splitted due
1687 (VS)	-	-	to dimer formation
-	1560 (S)	1540 (S)	$\nu_a(COO)$
1473 (M)	1472 (S)	1472 (M)	$\delta(CH_2)$ - Scissor; splitted due
1463 (S)	1462 (S)	1464 (S)	to crystal field
(1450-1187)†	present	present	$\delta(CH_2)$ - rock, wag, twist
1431 (M)	1431 (M)	1430 (M)	$\delta_a(CH_3)$
1411 (S)	1404 (M)	1404 (M)	$\delta(\alpha-CH_2)$
-	1397 (VS)	1397 (VS)	$\delta_s(COO)$
1311 (VS)	-	-	$\nu(C-OH)$
1298 (VS)	-	-	$\nu(C-OH)$
-	1179 (VS)	1175 (VS)	$\nu(C-OM^*)$
-	1111 (VS)	1115 (VS)	$\nu(C-OM^*)$
1102 (VW)	convoluted	convoluted	$\nu(C-C)$
940 (S)	-	-	$\delta(O-H)$
-	880 (S)	881 (S)	$\delta(O-M^*)$

ν: stretch; δ: deformation/bend; a: asymmetric; s: symmetric; M* = Cd, Co

VS: Very Strong; S: Strong; M: Medium; W: Weak; VW: Very Weak
† No. of band propagations of equally weak intensity.

cm^{-1} indicate orthorhombic packing [167]. However intensity of 1462 cm^{-1} peak is reduced in case of CoSt, as compared to the equally intense doublets in CdSt. This indicates possibility of deviation from true orthorhombic packing in CoSt. This difference in packing may lead to a shift in methyl stretching frequencies of hydrocarbon tail. To observe the methyl and methylene stretching frequencies, transmission-FTIR spectroscopy was carried out with 9ML films deposited on fused quartz, since Si is not transparent in the region of interest (Fig. 5.6(c)). The CH$_2$ asymmetric stretch (observed at 2917 cm^{-1}) showed a down shift of 2 cm^{-1} for CoSt compared to CdSt (observed at 2919 cm^{-1}), which is indicative of an increase in packing density [169], consistent with higher film coverage observed for CoSt.

NEXAFS study

Oxygen K-edge NEXAFS spectra of CoSt and CdSt films are shown in Fig. 5.7 (solid circles). Both spectra have been fit (solid line) with voigt function. Fitting parameters are shown in Table 5.7. The low energy sharper feature of the spectra arise due to π^* transitions whereas higher energy broader part denote σ^* transitions [147]. In CoSt, two σ^* peaks (3 and 4) and in CdSt three such peaks (5,6 and 7) are present. The low energy peak near 538 eV (present in both) is attributed to Si-O bonds coming from native SiO_2 layer of thickness $\sim 25\mathring{A}$ [170]. Higher energy peaks at 540.8 eV and 543.6 eV for CdSt correspond to O 1s$\rightarrow \sigma^*_{C-O}$ and O 1s$\rightarrow \sigma^*_{C=O}$ transitions respectively [171, 172]. Presence of both these peaks and also the π^* peak (marked as 1) at 532 eV corresponding to O 1s$\rightarrow \pi^*_{C=O}$ transition confirm presence of two asymmetric CO bonds in CdSt symmetric monolayer (SML) headgroup. As obtained from FTIR studies, there is only one Cd ion attached to two carboxylate groups in the SMLs (unidentate

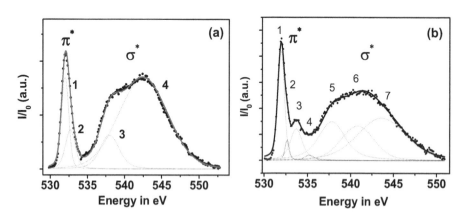

Figure 5.7: *O K-edge NEXAFS spectra of 9 ML (a) CoSt and (b) CdSt LB films. Experimental data (circles) have been fitted with voigt function (dotted lines). Solid lines are the composite fits.*

Table 5.7: Analysis of NEXAFS spectra
(x_c: Center; A: Amplitude; w_G: Gaussian width; w_L: Lorenzian width)

	Parameters of voigt fit			
CoSt	x_c	A	w_G	w_L
1	532.0	0.20	1.00	0.23
2	532.7	0.13	1.30	0.50
3	538.0	0.21	2.50	1.00
4	542.5	1.30	7.40	0.70
CdSt	x_c	A	w_G	w_L
1	531.9	0.22	0.80	0.18
2	532.7	0.03	0.40	0.30
3	533.7	0.10	1.00	1.00
4	535.2	0.02	0.80	0.80
5	537.8	0.37	2.3	2.2
6	540.8	0.35	4.0	1.0
7	543.6	1.65	5.5	2.00

[x_c, w_G and w_L are in eV, A is in a.u.]

coordination). This introduces an asymmetry in CO bond strength and orientation. The SML headgroups contain one C = O group along with another C-O group, latter attached to metal ion. However, due to delocalization, both bonds have partial π overlap, but the two CO bonds of the carboxylate group are not equivalent.

On the other hand, single CO sigma peak at 542.5 eV for CoSt denotes presence of symmetric CO bonds in CoSt headgroups. Energy of these CO bonds lie midway between that of C-O and C=O bonds in CdSt. Moreover, similar π^* peak near 532 eV is present. Thus two CO bonds of the carboxylate group present in CoSt bilayers are identical. These are consistent with bidentate bridging mode of metal ion coordination as obtained from FTIR study, as in this case symmetric CO bonds have nature intermediate between C-O and C=O bonds. It is also noted that π peaks in CdSt have smaller line widths compared to CoSt (Table 5.7), indicative of lower degree of delocalization consistent with asymmetric nature of bonds. Low intensity π^* peaks (2, 3 and 4) may be due to transition to Rydberg states and are not assigned [147]. It is to be noted that absence of Co-O or Cd-O bonds are due to high

Z-value of the metals since NEXAFS spectroscopy is applicable for detection of species with low Z values (Z<10) [147].

5.3.3   Morphology of Metal Stearate Multilayers

AFM study

AFM topographic images of both films are shown in Fig. 5.8. The CoSt film shows absence of "pinholes" (Fig. 5.8(a)). Line profile (Fig. 5.8 (a) inset-I) indicates a smooth surface except few isolated patches of height $\sim$127 Å. A closer scan, comparatively void of these isolated patches, was therefore analyzed (Fig. 5.8(b)). Bearing ratio obtained from AFM analysis (Fig. 5.8 (b) inset) shows no decrease of coverage with film thickness except at the top where there is sharp drop in coverage within a height of $\sim$30 Å comparable to ML thickness $\sim$27 Å. Distribution of CoSt film height measured from substrate (z_0) showed clear unimodality, with gaussian fit to peak as 127 Å, matching exactly with height of isolated patches. However, for region void of patches, this peak shifts to $\sim$28 Å (Fig. 5.8 (a) inset-II), consistent with ML thickness, clearly showing that pinholes present in CoSt films are strictly confined within topmost layer: an indication of very good film coverage along z_0. Estimated roughness is also quite low ($\sim$4 Å).

In sharp contrast to CoSt film, "pinholes" defects are present at CdSt film surface (Fig. 5.8 (c), (d)). Although defects actually grow in SML steps, line profile of CdSt (Fig. 5.8 (c) inset-I) drawn from (15×15 μm^2) image shows V-shaped pinholes due to convolution with AFM tip. Profile analysis shows presence of pinholes in each SML of CdSt, of different size and depth due to corresponding cumulative effect (average depth estimated as 248 Å). The bearing ratio (Fig. 5.8 (d) inset) shows a gradual decrease of coverage-area with z_0 - an indication of large number of pinholes. Moreover noticeable change in slope of CdSt bearing ratio at $\sim$150 Å, indicates enhanced decrease in film coverage above this height. Thus pinhole defects, although present from first ML, are greatly increased in number above the aforesaid height measured

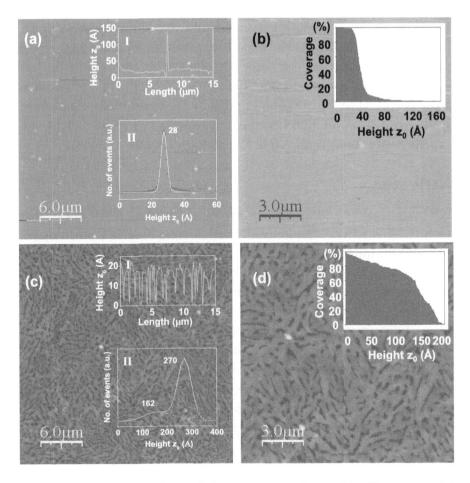

Figure 5.8: *AFM studies of the nine monolayer LB films of metal stearates.(a), (c) (30×30) μm^2 and (b), (d) (15×15) μm^2 topographic images; (a),(c) (insets-I) corresponding line profiles and (insets-II). Height (z_0) distribution data (circles) with corresponding Gaussian fits (line), (b), (d) (insets), corresponding bearing ratios. (a)–(b) correspond to CoSt and (c)-(d) correspond to CdSt films.*

from substrate. This is confirmed from bimodality in z_0 distribution of CdSt (Fig. 5.8 (c) inset-II). Gaussian fits to distribution peaks mark their positions as 162 Å and 270 Å. Higher z_0-value peak (at 270 Å) signifies total film thickness whereas lower peak (at 162 Å) corresponds to height above which number of pinholes increase greatly, agreeing with value obtained from bearing ratio. Average surface roughness including pinholes is quite high ($\sim$57 Å) for CdSt whereas defect-free regions are much less rough ($\sim$10 Å) showing little out-of plane fluctuation where BLs are present. However, it must be borne in mind that the height values obtained from AFM data include tip convolution effect. Precise height values were obtained from XRR analysis as discussed below.

XRR study

XRR data of the films are shown (open circles) in Fig. 5.9. For CoSt (Fig. 5.9 (a)) 7 Bragg peaks with strong Kiessig (interference) fringes and for CdSt (Fig. 5.9 (b)) 11 Bragg peaks with less prominent Kiessig fringes are obtained, showing CdSt to have lower coherence between air/film and film/substrate interfaces compared to CoSt. This is most probably due to presence of pinhole defects. Large number of Bragg peaks indicate both LB multilayers have good out-of-plane crystalline growth (comparatively better for CdSt). This apparently indicates no direct correlation between pinhole defects and out-of-plane crystallinity and suggests role of in-plane forces in defect formation. XRR data was analyzed using Parratt formalism [156], where each film is divided into layers of fixed thickness (d), average electron density (ρ) and interfacial roughness (σ), and used as fit parameters [50]. Solid lines in Fig. 5.9 (a) and (b) are best-fit curves. EDPs for the films with ρ as a function of film thickness z, where $z = 0$ is air, obtained from fits, are shown in respective insets. EDP of CdSt show gradual decrease in ρ with increase in SML number, from very first monolayer contacting substrate with H, T and G denoting head, tail and gap between two tails respectively. Pinholes reduce film coverage, giving rise to such EDP, consistent with AFM data. Air/film σ-value is 6 Å. EDP of CoSt show no such decrease of coverage to about $3\frac{1}{2}$ SMLs from substrate, and very small decrease in ρ for topmost SML. This indicates presence of few pinholes in CoSt primarily confined to top SML. Air/film σ-value is 6 Å for CoSt, consistent with AFM results. Thus, for CoSt film growth is more compact. However, height of pinholes in CoSt film obtained from

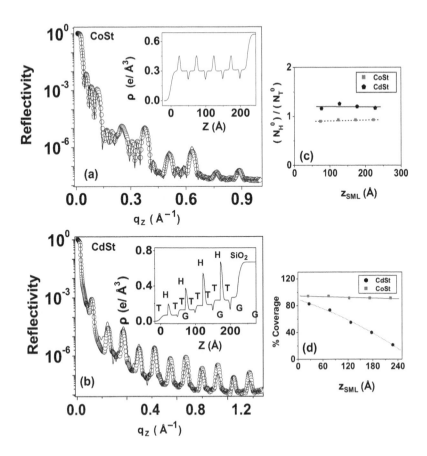

Figure 5.9: *(a), (b) x-ray reflectivity data (open circles) with corre-sponding fits (solid lines), of the deposited LB films (insets: corre-sponding EDPs (H, T, G denote head, tail and gap respectively)). (a) and (b) correspond to CoSt and CdSt films respectively. (c) calculated head-tail coverage ratio ($N_H^0 : N_T^0$) versus height (z_0) of the nine monolayer films. (d) calculated percentage coverage versus height (z_0) of the LB films.*

its EDP did not exactly match the AFM results. This is not sur-prising, as size of these pinholes are much smaller than those in CdSt film hindering the finite width of AFM tip to sense proper height variation.

Parameters of Good Growth

To quantify goodness of multilayer growth, assuming uniform molecular cross-sectional area A of 20 Å^2, we have calculated number of electrons N_H (N_T) in Head (Tail) for each SML as $\rho_H \times A \times d_H$ ($\rho_T \times A \times d_T$), where d_H (d_T) is headgroup (tail) thickness. Each N - value was normalized with corresponding bulk value N_b to account for porosity (pinholes). To eliminate decrease in ρ due to pinholes, the ratio $N_H^0 : N_T^0$ was obtained for each SML, where N_H^0 (N_T^0) is normalized value of N for head (tail) and plotted as a function of SML height (z_{SML}) for both films (Fig. 5.9 (c)). Slope χ of these curves is a measure of good out-of-plane crystalline growth, $\chi = 0$ denoting perfect multilayer stacking irrespective of presence of pinholes. χ-values obtained were $\sim 10^{-4}$ for both, showing excellent crystalline growth. Percentage film coverage, calculated from tails ($N_T \times 100$), versus z_{SML} (Fig. 5.9 (d)) show linear variation for CoSt but quadratic decay for CdSt, consistent with AFM results. Slope ($\chi' = 0.07$) of the linear fit indicates compact in-plane defect free coherent growth in CoSt with $\chi' \sim 0$ corresponding to perfectly coherent growth. Orientation of hydrocarbon tails were calculated from SML thicknesses obtained from reflectivity analysis. SML thickness for CdSt is 50.3 Å which corresponds to normal tail orientation to substrate whereas same for CoSt is 49.6 Å, corresponding to 9.6° tilt of tail from vertical.

5.4   On the Role of Metal–Headgroup Coordination

Infrared Reflection-Absorption Spectroscopy (IRAS) results on Cd-StA and Co-StA Langmuir monolayers on water surface, show that in presence of subphase metal ions, dissociated stearic acid forms unidentate, bidentate bridged and bidentate chelate coordinations. These Langmuir monolayers when transferred onto amorphous substrate, do not show all types of coordinations. Specific coordinations are selected depending upon metal ion type. our results show that the transferred monolayers contain unidentate

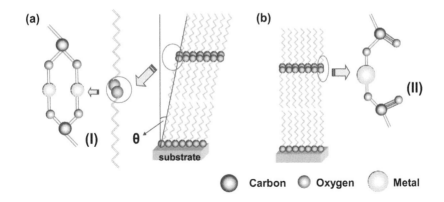

Figure 5.10: *Cartoon showing LB structure of (a) CoSt and (b) CdSt. θ represents the tilt angle of the hydrocarbon chain (θ = 9.6° and 0° for CoSt and CdSt respectively). The metal ion headgroup coordination is shown separately as (I) Bidentate bridging for CoSt and (II) Unidentate for CdSt (M = Cd or Co).*

and one of the bidentate coordinations of the carboxylate headgroup. Moreover, there is a strong dependence of the preferred headgroup coordination type on the metal ion. In other words, it is seen that Cd ions prefer unidentate coordination whereas Co ions prefer bidentate coordination. Although the exact mechanism of coordination selection during transfer (whether active or passive) is yet to be known, we propose that it is the action of particular and directed supramolecular interactions at the substrate surface that lead to the specific selections during transfer.

Again, the transferred multilayers of CoSt show bidentate bridging metal ion-carboxylate coordination, as opposed to unidentate coordination in CdSt. AFM and XRR results indicate large reduction of in-plane defects in CoSt multilayers compared to CdSt, along with excellent out-of-plane crystalline growth in both. we propose that the bidentate bridging coordination plays a key role in stacking of defect-free multilayers. Also CoSt films have higher hydrocarbon tail packing density; tails tilted at 9.6° to normal (Fig. 5.10). FTIR and NEXAFS studies clearly indicate underlying importance of headgroup structure in defining morphology of LB films. Structure and hence morphology of films is governed by metal-ion headgroup bonding.

Chapter 6

Headgroup Conformation and Film Structure

In the previous chapter, we have shown that morphology of transition metal stearate LB films depend on their headgroup coordination. More specifically, it is shown that "pinhole" defect-free LB multilayers are obtained for *bidentate bridged* headgroups in cobalt stearate. The next question that comes is the *sufficiency* of the bidentate bridged coordination in yielding near-perfect defect-free LB multilayers. We have answered this question by identifying specific molecular conformers created by processes of bulk reaction and reaction at air–water interface. We have also shown the correlation between the specific conformer and morphology of multilayers comprising molecules in that conformer. We have also discussed the significant role of supramolecular interactions at the air–water interface in deposition of highly reproducible coherent multilayers by choosing specific molecular conformer states.

In the following section, we have tested the sufficiency condition of the bidentate coordination in yielding defect-free multilayers. For this, we prepared thin films of cobalt stearate (CoSt) in the two distinct ways as given in Chapter 2. In the first, cobalt stearate prepared in bulk, was spread on water as the Langmuir monolayer from which the LB film was deposited. In this case head-tail ratio of amphiphiles in the monolayer is predetermined. However, for such preformed salts, exact molecular configuration at the air–water interface is not generally known. In the second way, the usual LB technique of vertical deposition with stearic acid as the monolayer on subphase water containing divalent cobalt ions, was employed. A comparative study of films deposited by both processes provided clues to arrive at the answer to the question of sufficiency for defect-free structures. We have used XRR and AFM to study the observed morphological changes of the LB films of cobalt stearate deposited in these two ways. These studies showed formation of a Volmer–Weber type monolayer but no multilayers for the preformed CoSt LB sample as compared to the excellent multilayers of CoSt films deposited at the air–water interface by the usual LB technique, in spite of both showing bidentate bridging type coordination of cobalt ions with the carboxylate group. We have attributed this difference to the existence of different headgroup conformers, as observed from FTIR studies. The preformed CoSt films had a higher energy "boat" conformation with linear O-Co-O linkage, whereas the ones formed at air–water interface formed a low energy conformer with a bent O-Co-O configuration (bond angle of 105°). Results discussed in this section support necessity of bidentate bridging coordination in multilayer deposition, but rejects its sufficiency by bringing out the crucial role played by supramolecular forces at air–water interface.

In the next section, we have investigated the role played by supramolecular forces at the air–water interface. By comparing the preformed and interfacial CoSt films, we have shown, using FTIR spectroscopy, that there is increased interaction between two methyl groups of adjacent hydrocarbon tails in case of the interfacial CoSt film, the interaction being of supramolecular nature. The origin of these supramolecular forces have been attributed to hyperconjugation. Moreover, we have indicated the dependence of the specific conformer formed at water surface on these supramolecular interactions. Methyl-methyl interactions observed in interfacial CoSt film suggest hierarchy of supramolecular chemistry at air–water interface in tuning the C-O-Co bond angle essential to satisfy the wetting condition with the substrate and subsequently form LB multilayers.

6.1 Preformed and Interfacial Langmuir–Blodgett Films

Thin Film Preparation

We have prepared thin films of CoSt in two distinct ways as given. In the first, cobalt stearate prepared in bulk (CoStb), was spread on water as the Langmuir monolayer. The thin films deposited from CoStb are named CoStp. In the second way, the usual LB technique of vertical deposition with stearic acid as the monolayer on subphase water containing divalent cobalt ions, was employed. Henceforth for clarity, these CoSt films will be called CoStn.

Formation of CoStb from stearic acid was carried out in the way described in Chapter 2 and confirmed by FTIR spectroscopic measurements [146] on bulk stearic acid (StA) and CoStb, drop cast on KBr substrates from chloroform solutions, in transmission mode. A chloroform solution of CoStb was spread in the Langmuir trough containing pure water and its surface pressure (π_s) vs specific molecular area (A) isotherm [5] was recorded by compressing the monolayer at a barrier speed of 3 mm/min. We deposited two CoStp films viz. CoStp-3 and CoStp-9, on hydrophilized fused quartz substrates by the LB technique, i.e., by three and nine subsequent vertical passages respectively of the substrate through the monolayer starting from water to air.

For interfacial cobalt stearate (CoStn) divalent cobalt ions were introduced in the Langmuir trough containing pure water through addition of cobalt chloride solution at pH = 6.0 as maintained by sodium bicarbonate. stearic acid monolayers were spread from a chloroform solution, were compressed at a barrier speed of 3 mm/min and the π_s-A isotherm was recorded. Again, we deposited two films, viz. CoStn-3 and CoStn-9 on fused quartz substrates by three and nine subsequent dips respectively, of the substrate through air–water interface starting from water, by the same LB technique. The films were studied using AFM, XRR and FTIR techniques with the same experimental parameters as in the previous chapter.

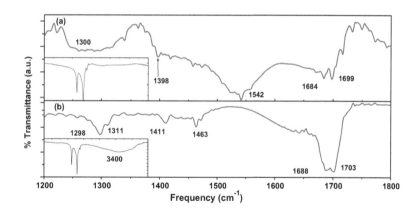

Figure 6.1: *FTIR spectra of (a) bulk cobalt stearate (CoStb) and (b) bulk stearic acid (StA). Methyl stretching region (2800 cm^{-1} - 3600 cm^{-1}) are shown in respective insets.*

Headgroup Coordination of Preformed cobalt stearate

Here we present FTIR results of bulk prepared cobalt stearate along with that of stearic acid. FTIR spectra of CoStb and StA are shown in Fig. 6.1. Assigned peaks are given in Table 6.1. Presence of strong bands corresponding to carboxylate asymmetric (1542 cm^{-1}) and symmetric (1398 cm^{-1}) stretch modes [166, 173] in the CoStb spectra (Fig. 6.1 (a)) indicate large conversion of fatty acid to its metal salt. Presence of very weak bands correspond-

Table 6.1: Infrared spectra of bulk cobalt stearate and stearic acid

Vibrational frequencies (cm^{-1}) of:		
CoStb	StA	Assigned to
-	3400 (S)	$\nu(O - H)$
1699 (M)	1703 (VS)	$\nu(C = O)$; split due
1684 (M)	1688 (VS)	to dimer formation
1542 (S)	-	$\nu_a(COO)$
1398 (VS)	-	$\nu_s(COO)$

ν: stretch; a: asymmetric; s: symmetric
VS: Very Strong; S: Strong; M: Medium

ing to C=O stretch (around 1700 cm^{-1}) and C-OH deformation (around 1300 cm^{-1}), and absence of OH stretch band (around 3400 cm^{-1}) in the sample indicate negligible amount of free acid and hydroxyl group [167], respectively, compared to strong acid peaks in StA spectra (Fig. 6.1 (b)).

In order to find coordination of metal ion with carboxylate group [161], difference in symmetric (ν_s) and asymmetric (ν_a) COO stretching frequencies ($\Delta\nu$) were measured. For CoStb, $\Delta\nu$ = 144 cm^{-1}, which corresponds to bidentate bridging coordination [174], having one metal ion per carboxylate group. As discussed in Chapter 5, CoStn LB multilayers form same bidentate bridging coordination of cobalt ion with COO group. *Thus identical metal ion carboxylate coordination exists in cobalt stearate samples prepared by the two routes.*

In general for bulk prepared samples, multiple metal ion coordinations can occur with sufficiently high subphase pH [175] and coordinations vary with chemical environment [176]. As such coordinations give well separated characteristic carboxylate stretch bands [176]. However, absence of other strong bands from 1550 cm^{-1} - 1600 cm^{-1} and from 1350 cm^{-1} - 1450 cm^{-1} in our bulk CoSt sample, clearly show that samples are quite free of such multiplicities and an almost pure bidentate bridging coordination is obtained. The fact that same coordination is obtained in the sample prepared at air–water interface points to two aspects. On one hand it shows that, bonding being same, the two systems CoStp and CoStn, can behave differently only by having different *conformations* , thereby bringing in the role of supramolecular forces in deciding their behavior. On the other hand it shows the selectivity of air–water interface towards a particular coordination of Co ions with carboxylate groups when the latter form a two-dimensional lattice at the interface, and Co ions associate with this lattice and distort it [177]. These aspects are discussed in the next sections.

6.2 Headgroup Conformation and Morphology

6.2.1 Monolayer Studies of Preformed and Interfacial Films

Figure 6.2 shows the π_s-A isotherms of CoStb monolayer on water (I) and StA monolayer with cobalt ions in water (II). The former show a lower collapse pressure compared to the latter. However, both show a region corresponding to the stable condensed phase, although at different pressure regions (18 mN/m for CoStb and 30 mN/m for StA+Co^{2+} system) and the films were deposited at these phases. Here, it must be mentioned that due to the low area per molecule and surface pressure values observed from the CoStp isotherm, existence of a true monolayer at the air–water interface may be questioned.

To investigate that, we deposited two more CoStp films under the same conditions. One was lifted horizontally (CoStp-H) by the

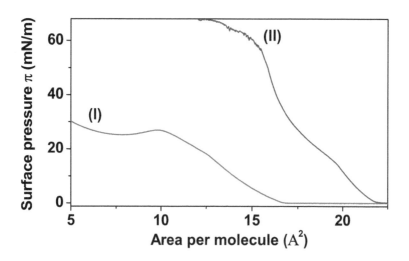

Figure 6.2: *Surface pressure π_s versus molecular area A isotherms of (I) CoStb monolayer on water and (II) StA monolayer with cobalt ions in water.*

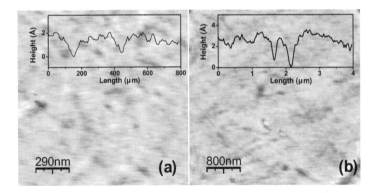

Figure 6.3: *AFM topographic images of (a) CoStp-H (1.5 μ m × 1.5 μ m) and (b) CoStp-V (4 μ m × 4 μ m) LB films with corresponding line profiles (insets).*

modified inverted Langmuir–Schaefer (MILS) technique [63] at a speed of 0.5 mm/min, and the other by lifting the substrate vertically (CoStp-V) through the air–water interface. The AFM topographic images of CoStp-H (Fig. 6.3 (a)) and CoStp-V (Fig. 6.3 (b)) show fairly uniform coverage along with some "pinhole" defects. Line profiles (insets) of both show the heights of the "pinholes" to be 25 Å, corresponding to monolayer heights. CoStp-H showed clearly that even at a lower area/molecule and surface pressure value, preformed cobalt stearate was present as a stable monolayer at the air–water interface and did not form multilayered structures. Furthermore, deposition of CoStp-V showed that this preformed system can be successfully transferred to a solid substrate as a true monolayer.

In terms of surface free energy per unit area σ, it is known from Bauer's criteria [82], the value of $\Delta\sigma = \sigma_o + \sigma_i - \sigma_s$ where suffixes o, i and s denote overlayer, interface and substrate respectively. $\Delta\sigma \leq 0$ denotes Frank van der Merwe (F-M) or "layer-by-layer" growth whereas $\Delta\sigma > 0$ gives Volmer–Weber (V-W) or "island-like" growth [81]. In this case, s is SiO_2 with $\sigma_s = 30.95$ mJ/m^2, obtained by dividing its Hamaker Constant value (reported as 6.4-6.6×10^{-20} J) [178] by a factor 2.1×10^{-21} m^{-2} [179]. The overlayer o is hydrocarbon tail of the monolayer, with $\sigma_o = 47.4$ mJ/m^2 for CoStb (17.7 mJ/m^2 for StA+Co^{2+} system), calculated from the basic relation $\pi_s = \sigma_w - \sigma_o$, where $\pi_s = 25.4$ mJ/m^2 (55.1 mJ/m^2) is measured surface pressure of CoStb (StA + Co^{2+}) Langmuir monolayer just before collapse and $\sigma_w = 72.8$ mJ/m^2 is that of pure water at 20°C. The reason for choosing collapse value of surface pressure for cobalt stearate is that it corresponds to the most closed packed condensed phase of monolayer on water. Therefore, for CoStb (StA + Co^{2+}), $\Delta\sigma = 16.5 + \sigma_i$ ($\Delta\sigma = $ -13.3 + σ_i), last term being interaction energy of cobalt ions with substrate, a small and often positive quantity.

The above analysis indicates that as long as interaction energies do not differ drastically, $\Delta\sigma$ values predict F-M growth for CoStn films and V-W growth for CoStp films. Thus, LB films of CoStp deposited at $\pi_s = 18$ mN/m will essentially show island-like formations resembling V-W growth mode. In Chapter 6, we showed that CoStn multilayers have F-M type growth. This will be discussed in details in Chapter 8. Thus the obtained morphologies of CoStp and CoStn are results of their behavior at air/water interface even with same salt having identical headgroup coordination deposited in the same way, illustrating importance of air–water interface in LB deposition. However it must be kept in mind that surface energy of overlayer was obtained from the Langmuir monolayer, which may differ from that on substrate, due to development of some amount of strain during transfer of monolayers onto substrate.

6.2.2 Morphology of Preformed and Interfacial Films

AFM study

AFM topographic images depicting surface morphology of CoStp-3 and CoStn-3 LB films are shown in Fig. 6.4. Presence of island-like formations are evident in CoStp-3 (Fig. 6.4 (a)) whereas CoStn-3 (Fig. 6.4 (b)) shows no such features. On the contrary, the latter shows a mesoscopically smooth morphology with negligible "pinhole" defects, consistent with the cobalt stearate multilayers

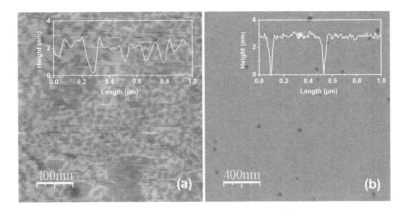

Figure 6.4: *AFM topographic images (2 μ m × 2 μ m) of (a) CoStp-3 and (b) CoStn-3 LB films with corresponding line profiles (insets).*

deposited by the same process as described in section (specify), on silicon substrates. Line profiles drawn through these images (Fig. 6.4 (a) and (b) (insets)) show heights of these islands (pinholes for CoStn-3) to be ~26 Å (30 Å for CoStn-3) which corresponds to a typical monolayer height. Estimated r.m.s. roughness is low (~ 5.8 Å for CoStp-3 and 1.9 Å for CoStn-3). Hence, AFM results show that in spite of bidentate bridging coordination, the CoStp-3 film did not show smooth, LB-like morphology.

It appears that CoStp contains growth defects, which stems from the fact that its surface energy on water at the deposited pressure characterizes V–W growth for this system as discussed for the monolayer. It must also be noted that AFM profile analysis indicated monolayer heights for both CoStp-3 and CoStn-3 films. For CoStn-3, with its smooth surface, this is an indication of very good film coverage along film thickness with pinholes strictly confined within topmost layer. For CoStp-3, this value should have been ~ 75 Å (height of a trilayer) considering presence of huge number of defects in the in-plane image. Although height values obtained from AFM data include tip convolution effect and may not give true height variations, yet the above result seemed puzzling. So to obtain precise out-of-plane information of deposited films, we carried out XRR measurements.

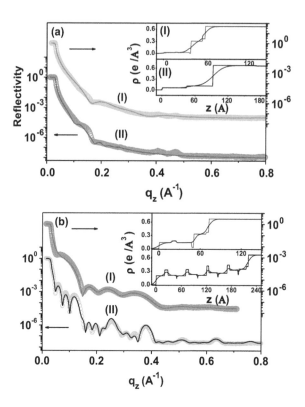

Figure 6.5: *X-Ray reflectivity data (open circles) with correspond-ing fits (solid lines) of (a) CoStp samples: (I) CoStp-3 and (II) CoStp-9, and (b) CoStn samples: (I) CoStn-3 and (II) CoStn-9, with their corresponding electron density profiles (EDPs) shown in respective insets.*

XRR study

XRR data of CoStp-3 and CoStp-9 films are shown (open circles) in Fig. 6.5 (a) (I) and (II) respectively. Both films indicate only a monolayer feature along with some "humps." The data has been fit using Parratt formalism with solid lines as best-fit curves. EDPs for films, obtained from fits, are shown insets, respectively. Values of the fit parameters of all films are given in Table 6.2. XRR data of CoStp-3 was fit by an AML of thickness 20.8 Å and a 39.8 Å thick SML of negligible electron density $\sim$0.02 electrons per Å^3. ML roughness estimated was 7 Å, consistent with AFM. CoStp-9

Table 6.2: Fitting parameters of x-ray reflectivity data for CoStp and CoStn samples

(h: head, t: tail, tt: tail-tail interface)

sample	layer	d(Å)	ρ(eÅ^{-3})	σ(Å)	sample	layer	d (Å)	ρ (eÅ^{-3})	σ(Å)
CoStp-3	t_3	17.0	0.02	4.0	CoStn-9	t_9	20.9	0.20	6.0
	h_2	3.8	0.03	0.5		h_8	7.8	0.33	4.7
	t_2	17.0	0.02	0.5		t_8	19.2	0.21	4.4
	tt	2.0	0.00	0.5		tt	1.3	0.14	1.5
	t_1	17.0	0.30	7.0		t_7	21.0	0.20	2.1
	h_1	3.8	0.36	0.0		h_6	6.8	0.34	4.7
	SiO$_2$	1×10^7	0.67	6.8		t_6	21.0	0.23	4.1
CoStp-9	Layer3	44.5	0.06	1.0		tt	1.6	0.15	0.0
	Layer2	48.3	0.11	8.6		t_5	21.0	0.21	1.0
	Layer1	80.0	0.68	15.5		h_4	5.0	0.42	4.5
	SiO$_2$	1×10^7	0.67	8.5		t_4	20.6	0.27	4.1
CoStn-3	t_3	18.4	0.14	6.1		tt	1.9	0.18	1.5
	h_2	6.9	0.18	1.7		t_3	21.0	0.26	2.3
	t_2	22.0	0.14	1.1		h_2	5.8	0.44	5.0
	tt	3.5	0.03	2.7		t_2	20.8	0.33	4.5
	t_1	18.2	0.33	4.2		tt	1.3	0.23	1.3
	h_1	3.2	0.37	1.8		t_1	18.8	0.35	3.0
	SiO$_2$	1×10^7	0.67	5.7		h_1	5.0	0.41	4.0
						SiO$_2$	1×10^7	0.67	8.0

data was fit with three arbitrary layers of thickness 80 Å, 48.3 Å and 44.5 Å having $\rho = 0.68$, 0.11 and 0.06 electrons per Å^3 respectively, i.e., *islands* varying from three to seven monolayers, suggesting an increase in coverage but *no LB multilayer* features were observed, consistent with AFM results.

XRR data of CoStn-3 and CoStn-9 films are shown (open circles) in Fig. 6.5 (b) (I) and (II) respectively. Both films show clear indication of *multilayer formation*. For CoStn-9, 5 Bragg peaks, with prominent Kiessig (interference) fringes are obtained. Prominent Kiessig fringes are an indication of coherence between air/film and film/substrate interfaces [135]. From the number of Bragg peaks it is evident that LB multilayers have good out-of-plane crystalline growth. Corresponding EDPs of CoStn (insets) show no considerable decrease in film coverage along z. This indicates presence of very few pinholes in films. Total film thickness is 72 Å for CoStn-3 and 221 Å for CoStn-9. Air/film interfacial widths (σ) are 6.1 Å (CoStn-3) and 6.0 Å (CoStn-9). Thus for CoStn, we obtained compact multilayer growth on fused quartz substrates.

XRR analysis clearly shows absence of LB multilayers in case of CoStp. Thus, island-like formations observed in AFM topographic image of CoStp are not "pinhole" defects arising during multilayer deposition. They merely represent a V-W type monolayer.

Thus, we can definitely say that metal ion coordination, although a necessary condition for defect-free morphology, as found in the previous section, is not sufficient to yield proper LB multilayers. In other words, bulk growth and interfacial growth produce cobalt stearate, that though are same as far as coordination is concerned, yet behave differently in supramolecular organization. As suggested before, the next step is to check their conformation, in particular, conformations of respective headgroups, where the metal is bonded to the organic amphiphile.

6.2.3   Determining Headgroup Conformations

FTIR study

In order to investigate that, we carried out a comparative FTIR study of both CoStp-9 and CoStn-9 which focussed on large amplitude vibrational modes associated with Co-bearing headgroups. The results are discussed in the following paragraphs. FTIR spectroscopy was carried out in Attenuated Total Reflectance (ATR) mode at 4.00 cm^{-1} resolution (Fig. 6.6 (a) and (b) for CoStp-9 and CoStn-9 respectively). ATR spectra was taken using Zinc Selenide plate, which has transmittance range above 650 cm^{-1}. So for the frequency range below 650 cm^{-1}, we took separate spectra in reflection mode, using silicon as substrate since fused quartz gave low signal in reflection mode. Assigned peaks of the spectra, after ATR and Krammers–Kronig corrections [146], are given in Table 6.3.

As discussed earlier, both films showed bidentate bridged Co-COO coordination. This requires two cobalt ions to be incorporated between two carboxylate groups, such that each cobalt ion is linked with two oxygen atoms, of different COO groups, on either side of it (Fig. 6.7(a)). Co-O stretch band lies in the range 665-703 cm^{-1} in cobalt oxide samples [180]. Since the COO stretching vibrations lie in the region 1400-1500 cm^{-1}, i.e., far

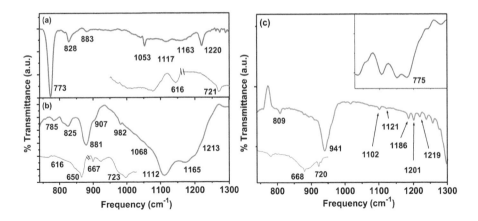

Figure 6.6: *FTIR spectra of (a) CoStp-9, (b) CoStn-9, and (c) stearic acid. FTIR spectra of CoStb (inset) showing presence of peak at 775 cm^{-1}.*

off from the O-Co-O stretch frequency range, O-Co-O and COO groups can be considered to be effectively decoupled and O-Co-O can be treated as isolated symmetric triatomic system giving rise to two stretching frequency modes, asymmetric (ν_a) and symmetric (ν_s), on either side of the diatomic X-Y stretching frequency, the asymmetric stretch being at a slightly higher value [163]. For a theoretical diatomic system Co-O, the stretching frequency comes out to be 640 cm^{-1}. Thus in both CoStp-9 and CoStn-9 samples, ν_a and ν_s are expected to lie above and below 640 cm^{-1} respectively. Comparing the two spectra and also with that of StA in this region (Fig. 6.6 (c)), we found that CoStp-9 (CoStn-9) has two distinct peaks at 773 cm^{-1} (650 cm^{-1}) and 616 cm^{-1}(616 cm^{-1}), which we have tentatively assigned to ν_a and ν_s respectively. It is worthwhile mentioning that presence of the peak at 775 *cm*$^{-1}$ in CoStb (Fig. 6.6 (c)(inset)) indicates that headgroup conformers of CoStp-9 and CoStb are similar, and hence presumably remain unchanged at air/water interface. In other words, bulk CoSt sample formed during chemical reaction exists in a particular headgroup conformer state and it remains in that conformer state throughout.

Table 6.3: Infrared spectra of CoStn and CoStp samples

Vibrational frequencies (cm^{-1}) of:			
CoStn	CoStp	StA	Assigned to
-	1220(M)	1219(W)	$\delta(CH_2)$ - wag
1213(C)	-	1201(W)	$\delta(CH_2)$ - wag
-	-	1186(W)	$\delta(CH_2)$ - wag
1165(VS)	1163(W)	-	$\delta(CH_2)$ - wag and/or $\nu(C-O-Co)$ stretch
-	-	1121(VW)	$\nu(C-C)$ - skeletal
1112(VS)	1117(W)	-	$\nu(C-C)$ - skeletal and/or $\nu(C-O-Co)$ stretch
1068(VS)	1100(VW)	1102(W)	$\nu(C-C)$ - skeletal
982(M)	1053(M)	1074(VW)	$\nu(C-C)$ - skeletal
881(VS)	883(W)	880 (VW)	skeletal plane and/or $\delta(C-O-Co)$ deformation
-	773(VS)	-	$\nu_a(O-Co-O)$
650(VS)	-	-	$\nu_a(O-Co-O)$
616(M)	616(S)	-	$\nu_s(O-Co-O)$

[VS: Very Strong; S: Strong; M: Medium; W: Weak; VW: Very Weak; C: Convoluted; ν: Stretch; δ: Deformation/Bend (includes scissor, rock, wag, twist); a: asymmetric; s: symmetric]

For a symmetric triatomic molecule X-Y-X, the asymmetric (ν_a) and symmetric (ν_s) stretching frequencies are given by the relations,

$$\nu_{a,s} = (2\pi c)^{-1} k^{1/2} [m^{-1} + M^{-1}(1 \mp cos2\alpha)]^{1/2} \qquad (6.1)$$

where m and M are masses of X and Y atoms respectively, 2α is bond angle, c is velocity of light and k is force constant of X-Y bond [162]. Value of k was calculated from the empirical relation for approximate force constant of a single bond between atoms X and Y as $k_{XY} = 7.20 \times 10^2 (Z_X Z_Y)(n_X n_Y)^{-3}$ N/m, where Z and n are atomic number, and principal quantum number of valence electrons, respectively [164]. Theoretical k value came out as $(3.04 \times 10^2$ N/m). From the above two simultaneous equations, we determined experimental values of k and α of CoStp-9 and CoStn-9 (Table 6.4), using observed ν_a and ν_s values for O-Co-O stretch mode of the samples. No unique solution for k and α was

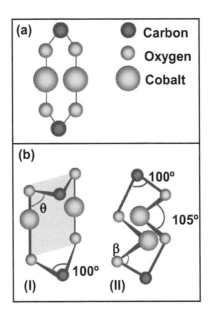

Figure 6.7: *Cartoon showing (a) the bidentate bridged headgroup along with (b) the headgroup conformations of (I) CoStp and (II) CoStn samples.*

obtained for CoStp-9 but, when k versus α curves obtained from both equations are plotted (Fig. 6.8 (a)), k values of both curves approached each other as α approached 90° and these values were chosen as the solutions. The value of $\Delta\nu = (\nu_a - \nu_s)$, which varies as 2α, was also calculated. Theoretical $\Delta\nu$ values were calculated for all values of α from 0° to 90° (Fig. 6.8 (c)), and a plot of $\Delta\nu_{Th}$ vs α (Fig. 6.8 (c) inset) was obtained. The values are presented in Table 6.4. As seen from table 6.4, observed and calculated k and $\Delta\nu$ values of CoStn-9 are in good agreement. Most importantly, α values show that O-Co-O linkage is linear in CoStp-9 but nonlinear for CoStn-9 with bond angle of 105°. For transition metals, 3d-2p hybridization affects this bond angle [181], with the latter decreasing with an increase in former, hinting at existence of different hybridization in each case. Difference in bond angles point out that although both headgroups are bidentate bridged, *they definitely exist as different conformers [182] with different*

Table 6.4: Force constants and bond angles from FTIR analysis

Triatomic system	Sample	ν_a (cm^{-1})	ν_s (cm^{-1})	$\Delta\nu$ (cm^{-1})	$k \times 10^2$ (N/m)	α (deg)	$k_{Th} \times 10^2$ (N/m)	$\Delta\nu_{Th}$ (cm^{-1})
O-Co-O linkage	CoStp-9	773	616	157	3.61	90.0	3.04	148
	CoStn-9	650	616	34	2.98	52.3	3.04	36
O-C-O linkage	CoStp-9	1542	1398	144	8.77	50.0	8.67	145
	CoStn-9	1541	1397	144	8.76	49.9	8.67	145

hybridization states, with bulk samples having higher energy, since it has higher ν_a value.

Carboxylate linkage in both samples are investigated next. From carboxylate stretch bands mentioned before, we carried out similar calculations for determination of experimental values of force constant k and bond angle α (labeled as α_C in Fig. 6.6 for clarity). Results are given in Table 6.4. However, since carboxylate group in bidentate bridging coordination behaves similar to a carboxylate free ion [90], and since the latter exhibits a CO bonding intermediate between a pure single and double bond due to ex-

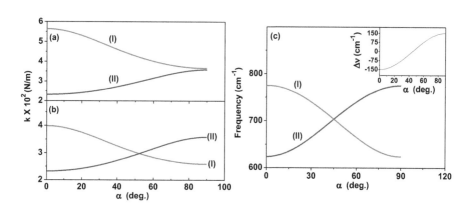

Figure 6.8: *Variation of force constant k versus bond angle α for O-Co-O bond in (a) CoStp and (b) CoStn samples. Curves (I) and (II) correspond to asymmetric and symmetric O-Co-O stretching frequencies. (c) theoretical variation of (I) ν_a, (II) ν_s and (inset) $\Delta\nu_{Th}$ with bond angle α for a O-Co-O triatomic system.*

istence of the resonance hybrid structure [89], theoretical k value was computed using the expression of force constant for multiple bond as $k_N = k_{XY} N(r_1/r_N)$ where k_{XY} is force constant for X-Y single bond, N is bond order, r_1 and r_N are single and multiple bond lengths [164]. Taking $N = 1.5$, $r_1 = 1.36$ Å and $r_N = 1.27$ Å, k_{Th} came out to be 8.67×10^2 N/m. Comparing experimental values for COO (which are in good agreement with theoretical ones), we found that carboxylate group was similar in both samples.

From the above analysis, possibility of existence of planar metal–carboxylate headgroup geometries in both CoStp and CoStn films can be ruled out, since a planar geometry would lead to greater delocalization of π bonds of COO leading to coupling between O-Co-O and COO systems, which is inconsistent with an isolated triatomic system.

Taking into consideration the fact that CoStb forms a stable monolayer upto 26 mN/m, as seen from its $\pi_s - A$ isotherm, we assume that the preferred conformer state is a "boat" (Fig. 6.7 (b)(I)), with two O-Co-O groups forming its base and two carboxylate groups with their hydrocarbon tails being on same side, i.e., away from water at the air/water interface with a fixed value of C-O-Co angle θ. Taking into account covalent radii of Co and O, area of the base of the boat was calculated as 15.75 Å^2, which was very close to the value (12.5 Å^2) of area/molecule of CoStp at 18 mN/m pressure as obtained from its isotherm. Although it is difficult to quote the exact value of θ from the above data, nevertheless it can be definitely said that θ remains frozen at the air/water interface and also in CoStp samples, as FTIR spectra of CoStb and CoStp samples show similar spectral features, suggesting a rigid headgroup structure.

Similarly, from structural data obtained above, and taking into account covalent radii of Co ions, a tentative headgroup conformer is constructed for CoStn (Fig. 6.7 (b) (II)). As seen from the cartoon, the headgroup is a twisted conformer with O-Co-O plane inclined at some angle β (say) with plane of four O atoms of two COO groups. However, it is difficult to quote the exact value of β at this stage.

It must be mentioned here that evaluation of force constants and bond angles for O-Co-O and O-C-O linkages were done considering harmonic vibration potentials neglecting corrections due to anharmonicity, which might be in question considering suggested molecular arrangement, and transitions between vibrational states, which might happen during film transfer. However, due to considerable frequency separation between O-Co-O and O-C-O linkages, they are effectively decoupled systems owing to which anharmonicity can be neglected, even for the suggested two headgroup conformers. Also, transition between vibrational states can be neglected in above calculation, since we am comparing FTIR spectra of LB films after transfer on to substrates.

Above results say that the two modes of preparation yield two *different molecules*. Though this difference is in the conformation of headgroups, it means that they are in different hybridization states. This answers the first question raised earlier in this chapter and shows the importance of water surface in facilitating growth of the alternate conformer. We will show in the next section how supramolecular forces play a stronger role at the water surface in selecting this conformer.

6.3   Headgroup Conformation and Supramolecular Interactions

The weak intermolecular forces in organic i.e., molecular crystals lead to incoherence in growth of these crystals and consequent lack of reproducibility. Controlling supramolecular interactions and, in particular, enhancing them by selecting specific molecular configurations should play a key role in growing functional organic crystals with a high degree of reproducibility. For that, the specific interconnections between supramolecular forces, molecular conformations and multilayer deposition need to be elucidated. In this section, we have studied the relation between molecular configuration and supramolecular forces as well as the significant role of the latter in deposition of highly reproducible coherent LB multilayers of cobalt stearate.

6.3.1 Origin of Supramolecular Interactions

In this subsection, we discuss the origin of supramolecular forces and the significant role of supramolecular interactions in forming LB multilayers. For this, we have carried out a comparative study of CoStn films with those of preformed bulk cobalt stearate (CoStp), as in the former, highly reproducible coherent multilayers are obtained in contrast to the V-W type monolayers obtained in case of the latter. Information about hydrocarbon tail-tail interactions have been obtained from FTIR spectroscopy by studying methyl stretching frequency regions (2800 cm^{-1} to 2950 cm^{-1}) of CoStn multilayers along with those of stearic acid and bulk CoStp using transmission mode. The metal-ion headgroup coordination (around 700 cm^{-1}) of the CoStp film was also studied.

The methyl asymmetric and symmetric stretching frequency regions of these three samples are shown in Fig. 6.9. In contrast to single peaks for stearic acid (Fig. 6.9 (a), (d)), the cobalt

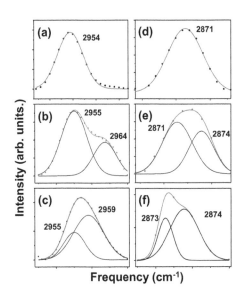

Figure 6.9: *FTIR spectra of (a)-(c) methyl asymmetric and (d)-(f) symmetric stretching frequencies of (a), (d), bulk stearic acid, (b), (e) preformed cobalt stearate and (c), (f) CoSt LB multilayers.*

stearate samples show presence of two peaks in both the asymmetric (Fig. 6.9 (b), (c)) and symmetric (Fig. 6.9 (e), (f)) methyl stretch modes. In general when two methyl tops interact they give two conformer states, viz. the staggered and eclipsed states [182], which coexist. Both of them have different symmetric and asymmetric stretch modes, although differing slightly in energy. Thus, the existence of two stretch modes, both symmetric and asymmetric, in cobalt stearate is consistent with the coexistence of both conformers in the samples. In contrast, for stearic acid, these two states remain unresolved. The possible reason behind such observation may be that stearic acid headgroup region has two distinct COOH groups connected by weak hydrogen bonding. This situation leads to comparatively freely rotating methyl tops.

The higher frequency mode for the asymmetric stretch is downshifted by 5 cm^{-1} in CoStn (Fig. 6.9 (c)) compared to bulk CoStp (Fig. 6.9 (b)), indicating weakening of C-H bonds in methyl groups for CoStn. This redshift in frequency corresponds to an energy $\sim$ 0.09 kcal/mol. Moreover, there is an observed increase in intensity of the aforesaid peak at 2964 cm^{-1} (Fig.5 (b)) by a considerable amount (peak at 2959 cm^{-1} in Fig. 6.9 (c)) relative to the peak at 2955 cm^{-1} indicating coupling of two methyl tops attached to hydrocarbon tails of adjacent SMLs via some supramolecular interactions . Now, a redshift of C-H stretching frequencies, attributed to lengthening of C-H bonds, is a signature for formation of a supramolecular bond such as hydrogen bond [183]. In fact, the above notion is a generalization observed in case of any X-H..Y supramolecularly bonded complex [184] where X-H bond lengthening occurs due to hyperconjugative interaction $\sim$ 0.1-10 kcal/mol; the underlying theory extending to understanding of intramolecular stereoelectronic effects on molecular geometry [185]. Moreover, importance of intramolecular vicinal hyperconjugation [186] in choosing the minimum energy configuration state of a molecule having two methyl tops has been established. The above discussion hints at the possibility of some very weak *intermolecular* hyperconjugative effects to be the reason behind the redshift in C-H frequencies in this case. Possibility of dipolar interactions ($\sim$ 2.5 kcal/mol - 25 kcal/mol) or Van der Waals forces ($\sim$ 0.24 kcal/mol) are less probable as the order of interaction energy falls below these typical values. However, establishing the exact nature of the supramolecular interaction is yet to be established.

6.3.2 Conformation Selection by Supramolecular Forces

It has been shown that in LB multilayers comprising two-tailed amphiphiles, molecular conformation allows only tails to take part in supramolecular and interlayer coupling [62]. In particular, interlayer hydrocarbon tail-tail interactions take place predominantly via methyl tops [61]. Origin of such methyl-methyl interaction has been attributed to hyperconjugation, as discussed in the previous subsection. In this subsection, our focus is to elucidate effect of such interactions on molecular structure, using FTIR spectroscopy.

Interaction between two methyl tops causes linking of adjacent hydrocarbon chains which may in turn affect C-C skeletal vibrations [187], another very important large amplitude vibration mode associated both with conformation and supramolecular forces. Ordinarily, C-C stretch modes are IR weak [161] such that most of the 17 C-C modes in stearic acid [164] remain undetected. Simple calculations show that increasing the number of C atoms from 18 (one stearic chain) to 36 (two coupled stearic chains) causes each C-C stretch mode to be doubly degenerate, thereby increasing their intensity. Also, hydrocarbon tail linkage would result in coupling of CH_2 wag modes of both chains [163], causing their intensity enhancement. Thus, in our case, increased intensity of both C-C skeletal and CH_2 wag modes are expected in CoStn compared to CoStp and StA.

FTIR spectra of CoStp-9 (Fig. 6.6 (a)) and CoStn-9 (Fig. 6.6 (b)), showing regions of C-C skeletal vibrations (1000-1200 cm^{-1}) in carboxylate salts [166], are clearly different. Peaks at 1112 cm^{-1}, 1068 cm^{-1} and 982 cm^{-1} present in CoStn-9, attributed to C-C skeletal modes (Table 6.3), are enhanced in intensity as well as frequency downshifted compared to those in CoStp-9 and StA (in an all-*trans* C form) [188]. The increased intensity of these C-C modes in CoStn are definite indications of hydrocarbon chain coupling. Again frequencies at 1165 cm^{-1} and 1213 cm^{-1} in CoStn-9, attributed to methylene wag, are enhanced in intensity compared to CoStp and StA, consistent with our presumption. However, a second possibility remains that the peaks at 1112 cm^{-1} and 1165 cm^{-1}, are indeed absent in StA, in which case, their origin may be attributed to metal–carboxylate interactions in CoSt samples. Even then, CoStn spectra shows clear signatures of enhanced *ordering* of tails due to supramolecular interaction, and the fact that corresponding peaks follow an ascending intensity scale from StA to CoStp to CoStn indicate that strength of supramolecular forces follow the same scale. This is consistent with interlayer coupling through the end methyl tops, where the tops are *staggered* to each other, a configuration that has been found to promote hyperconjugation within a molecule [186].

CoStn spectra show a prominent band at 881 cm^{-1} that appears as very weak in spectra of CoStp and StA. This band, also present in the spectra of cadmium stearate LB film at 880 cm^{-1} (Fig. 5.6, is characteristic of LB multilayers. It was tentatively assigned to the metal–carboxylate deformation frequency. Although the C-O-Co deformation is different in the CoStp and CoStn samples, which supports the above observation, yet the presence of even a very weak peak near this frequency in CoStp raises some amount of uncertainty in this assignment. This is because the methyl rocking frequency lies close to this value as observed in C form of stearic acid [188]. The StA spectra shows presence of two very weak peaks at 893 cm^{-1} and 880 cm^{-1}, attributed to methyl rock. Thus a second, and probably more justified possibility is that the peak at 881 cm^{-1} indeed corresponds to methyl rocking mode, with an increased intensity in case of normal LB multilayers. This is because interlayer supramolecular coupling between adjacent methyl tops freezes free rotation of C-C skeletal planes of adjacent hydrocarbon chains in CoStn multilayer, compared to that of StA, thereby enhancing the methyl rock mode. Although a detailed analysis is necessary for precise assignment of this band, it can be said that enhancement of this peak, along with the enhancement in skeletal vibration, suggests the enhancement of interlayer forces.

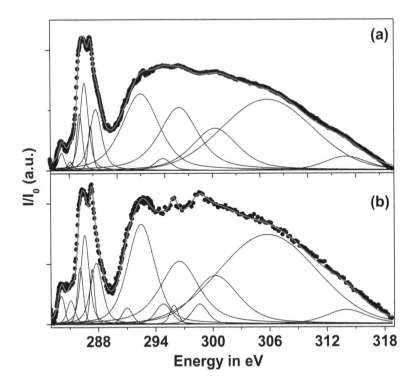

Figure 6.10: *C K-edge NEXAFS spectra of 1 ML CoStn film for (a) normal and (b) grazing angles of incidence. Experimental data (circles) have been fitted with voigt function (dotted lines). Solid lines are the composite fits.*

Carbon K edge NEXAFS spectra (open circles) of 1 monolayer CoStn deposited on silicon at normal and grazing (20° to substrate) angles of incidence (Fig. 6.10 (a) and (b) respectively) give supportive evidence regarding the hydrocarbon tail orientation w.r.t. substrate [147]. The spectra is fitted with voigt function (solid line). Position and amplitude of fitted peaks are given in Table 6.5. Although individual peak assignments are not done, the spectra is broadly classified as having prominent features around 287 eV, 295 eV and 305 eV, as observed in LB monolayers of fatty acid salts [189]. The two sharp features around 287 eV, characteristic of hydrocarbon chains (fitted by peaks 1 to 6 in

Table 6.5: Results of NEXAFS analysis of 1 ML CoStn film

Peak	x_c	A_G	A_N
1	284.1	0.08	0.06
2	285.1	0.10	0.04
3	286.1	0.17	0.17
4	286.6	0.42	0.41
5	287.4	0.14	0.11
6	287.8	0.42	0.42
7	292.5	1.43	1.55
8	294.8	0.17	0.10
9	296.5	1.40	1.40
10	300.3	1.20	1.04
11	305.8	3.80	2.80
12	314.0	0.30	0.29
13	291.0	0.09	-
14	295.9	0.08	-
15	298.6	0.17	-

x_c *(eV): position, A_N (A_G) (arb. units): amplitude at normal (grazing) incidence*

Table 6.5) are usually assigned to C 1s $\rightarrow \sigma^*_{C-H}$ transitions [189]. However, the same have been attributed to C 1s $\rightarrow$ Rydberg transitions [190] and mixed Rydberg/valence orbitals (of antibonding σ^*_{C-H}) transitions [191], and so there remains ambiguity regarding their assignment. Features around 295 eV and 305 eV are shape resonances typically assigned to C 1s $\rightarrow \sigma^*_{C-C}$ transitions and C 1s $\rightarrow \sigma^*_{C-C'}$ transitions [189]. The amplitudes of all shape resonances (except peak at 292.5 eV), are found to increase in the grazing incidence spectra. From these results, we suggest that the hydrocarbon tails are oriented more towards the substrate normal. The C 1s $\rightarrow \sigma^*_{C-O}$ and C 1s $\rightarrow \sigma^*_{C=O}$ transitions, typically observed at 296.5 eV and 303.5 eV in formic acid spectra respectively, are generally weak compared to the C-C transitions as amplitude of carboxylate resonances are about 5 % of the same for hydrocarbon resonances. Although our spectra show the presence of weak peaks, which may be due to carboxylate resonances, it is not possible to precisely assign them at this point.

Thus we find that the molecules of cobalt stearate formed at the air–water interface show considerably higher supramolecular bonding than those formed by the reaction in bulk. The enhancement in supramolecular interaction comes from more flexibility in orientation of the chains, which in turn, is brought about by the flexibility in the C-O-Co angle or θ. In case of CoStn deposition, θ is not fixed beforehand as the formation of cobalt stearate occurs during the deposition process at the air/water interface itself and the "twisted" conformer is chosen by the supramolecular forces to culminate in the FM growth mode with low free energy. This is manifested in the microscopic scale as "perfect" LB multilayers as shown in Chapter 5. In contrast, for bulk reaction, CoSt with the "boat" conformer as headgroup (CoStb and CoStp) is formed. As we have shown, this is a rigid structure that remains unchanged throughout. Hence, θ is frozen leading to the orientational rigidity of the chains and lessening the role of supramolecular forces. This is manifested mesoscopically by the fact that LB multilayers cannot be deposited on this rigid structure. On repeated dipping of the substrate, increase in coverage is observed in CoStp, but no stable multilayers are formed.

6.4 On the Role of Headgroup Conformation and Supramolecular Forces

We have shown through Fourier Transform Infrared (FTIR) Spectroscopy that, cobalt stearate prepared in bulk and at air/water interface has identical metal ion-carboxylate coordination, viz. the bidentate bridged configuration. However, both X-Ray reflectivity (XRR) and Atomic force microscopy (AFM) results show that preformed cobalt stearate film (CoStp) is deposited as a V-W type monolayer in contrast to the ones at air/water interface (CoStn), the latter having multilayers with good out of plane crystallinity and near perfect defect free morphology. We have associated this difference in their morphology with different conformations of their Co-bearing headgroups. FTIR study reveals that CoStp headgroup (and also CoStb) exist as a "boat" conformer with linear O-Co-O linkage, whereas CoStn forms a lower energy "twisted" conformer with a bent O-Co-O configuration having a bond angle of 105°. The flexible headgroup structure in CoStn allows weak supramolecular interactions , of hyperconjugative origin, between methyl tops at chain-ends. These weak interactions couple the hydrocarbon chains end-to-end and order their skeletal planes, thereby reducing free energy and leading to excellent Langmuir–Blodgett multilayers. On the other hand, the headgroup structure in CoStp (or CoStb) is very rigid and inhibits supramolecular bonding. As a result, multilayer growth cannot take place. Such adverse effect of monolayer rigidity on multilayer formation has been observed elsewhere [192] and our results thus support a general notion in this regard.

The importance of our finding lies in demonstrating a definite role of the air–water interface in selecting bond coordination and conformation, on one hand and relating morphology in the microscopic length scale to bonding molecules via molecular conformations that allow or inhibit specific supramolecular forces and large amplitude motions, on the other. The first point is presumably connected to the observation of hydronium ion enrichment at water surface [193], while the second maybe utilized to tune self-organization of complex structures by the low-energy modes of large amplitude motions.

Chapter 7

Evolution of Film Morphology

Work discussed in the last two chapters relate to the dependence of morphology on headgroup coordination for two divalent metal stearates, viz. cobalt and cadmium. Although our results establish the association of defect-free morphology to the "twisted" conformer of bidentate bridged coordination in cobalt stearate yet they do not give systematics of such dependence. In this chapter, we have tried to study the growth mechanism of different divalent metal-bearing amphiphiles and tried to classify them under the general heteroepitaxial growth modes . Further, we have tried to elucidate the underlying reason behind dependence of morphology and on the headgroup structure. To this end one, three and nine MLs each of cadmium stearate (CdSt), zinc stearate (ZnSt), cobalt stearate (CoSt) and manganous stearate (MnSt) were deposited on hydrophilic Si(100) substrate by LB technique, as described in previous chapters. All films were checked for reproducibility. AFM of the films was performed in tapping mode with a silicon cantilever. Scans were performed over several regions of the films for different scan areas. XRR and FTIR spectroscopy in ATR mode were carried out.

7.1   Morphological Evolution and Growth Modes

In this section, we have studied the morphological evolution of different divalent metal-bearing amphiphiles, viz. Cd, Zn, Co and Mn by comparing their monolayers, trilayers and multilayers, using AFM and XRR techniques. We have studied the morphological change w.r.t. (I) layer number and (II) atomic number of metal. Results show a strong dependence of morphology on atomic number of metal. Further, we obtained information about their out-of-plane growth and classified them under the three heteroepitaxial growth modes. Each metal stearate film exhibited different growth mode, with Mn and Co bearing films showing *Frank Van der Merwe* type growth, Zn bearing films showing *Stranski–Krastanow* type growth and Cd bearing ones showing *Volmer–Weber* type growth.

7.1.1   Atomic Force Microscopy Results

AFM topographic images of 1ML, 3ML and 9ML films of MnSt and CoSt with their corresponding height distributions (insets) are shown in Figure 7.1. The same for ZnSt and CdSt are shown in Figures 7.2 and 7.3 respectively. The AFM height distribution data (solid circles) have been fit with Gaussian function (solid line) such that all height values are obtained from the fitted peak positions. We have studied the morphological evolution of these films w.r.t. (I) layer number and (II) atomic number of metal. They are discussed one after the other as follows:

Evolution with Layer Number

In order to study the morphological evolution with layer number for each MSt film, morphology of 1ML film is compared with those of its 3ML and 9ML counterparts. For example, 1 ML films of MnSt (Fig. 7.1 (a)) and CoSt (Fig. 7.1 (d)) show unimodal distributions with peaks at 12 Å and 14 Å respectively, denoting true monolayer features, with uniform coverage (calculated r.m.s.

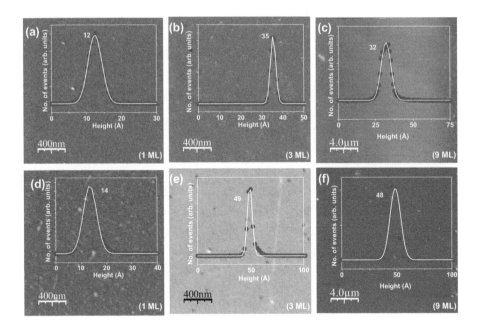

Figure 7.1: *AFM topographic images of MnSt and CoSt films show-ing Frank Van der Merwe type growth. In the figure, (a),(d) repre-sent 1ML (2 × 2 μ m²), (b),(e) represent 3ML (2 × 2 μ m²) and (c),(f) represent 9ML (20 × 20 μ m²) films of (a)-(c): MnSt; (d)-(f): CoSt LB films with corresponding height distributions (inset). In all insets, height distribution data (plotted as dots) is fitted by Gaussian function (solid line).*

roughnesses are 2.0 Å and 3.5 Å respectively). Again, 3ML films of MnSt (Fig. 7.1 (b)) and CoSt (Fig. 7.1 (e)) show formation of defect-free smooth SML (r.m.s. roughness of 5 Å and 2 Å re-spectively) on top of the first AML. Unimodal distributions show very few pinholes of SML thickness (35 Å for MnSt and 49 Å for CoSt). The 9 ML counterparts of MnSt (Fig. 7.1(c)) and CoSt (Fig. 7.1(f)) resemble morphologies of their 1ML and 3ML coun-terparts, having unimodal distribution peaks at 32 Å and 48 Å re-spectively (r.m.s. roughnesses are 4.2 Å and 7.1 Å respectively). Thus *surface morphology of MnSt and CoSt do not change with layer number* up to at least four SMLs after the AML.

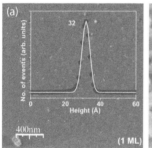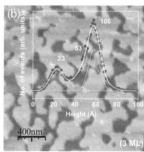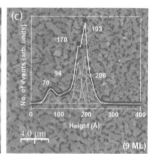

Figure 7.2: *AFM topographic images of ZnSt films showing Stranski–Krastanov type growth. In the figure, (a), (b) and (c) represent 1ML ($2 \times 2 \mu m^2$), 3ML ($2 \times 2 \mu m^2$) and 9ML ($20 \times 20 \mu m^2$) films respectively with their corresponding height distributions (inset). Height distribution data (plotted as dots) is fitted by Gaussian function (solid line), where thick line represents composite fit and thin lines represent individual peak fits.*

On the other hand, surface morphology of 1ML ZnSt (Fig. 7.2(a)), which also shows true monolayer feature (unimodal distribution peak at 32 Å and r.m.s. roughness 3.8 Å) is found to change with deposition of a SML on top of the AML as is evident from the 3ML ZnSt film (Fig. 7.2 (b)); morphology of the latter shows development of pinholes (height distribution showing islands of heights 23 Å, 53 Å and 106 Å; with r.m.s. roughness of 16 Å). Surface morphology of 9 ML ZnSt film (Fig. 7.2 (c)) resembles that of its 3ML counterpart and has a multimodal distribution of different heights (70 Å, 94 Å , 170 Å , 193 Å and 206 Å). So, *surface morphology of ZnSt changes with layer number* such that island formation commences with deposition of first SML on the AML and continues up to at least next three SMLs.

Surface morphology of 1ML CdSt film (Fig. 7.3 (a)) is found to differ drastically from the rest. In sharp contrast to the observed unimodal distributions of the 1ML depositions of the other three MSt films, 1ML CdSt comprises islands of various heights (62 Å, 92 Å and 134 Å) as is evident from its multimodal distribution. R.m.s. roughness estimated was 36 Å. Again, morphology of 3ML CdSt film (Fig. 7.3 (b)) shows similar distribution with domain-

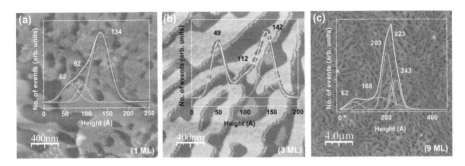

Figure 7.3: *AFM topographic images of CdSt films showing Volmer–Weber type growth. In the figure, (a), (b) and (c) represent 1ML (2 × 2 μ m²), 3ML (2 × 2 μ m²) and 9ML (20 × 20 μ m²) films respectively with their corresponding height distributions (inset). Height distribution data (plotted as dots) is fitted by Gaussian function (solid line), where thick line represents composite fit and thin lines represent individual peak fits.*

like growth of various heights (49 Å, 112 Å and 142 Å) separated by dark ridges (r.m.s. roughness was 42 Å). Peak of height 49 Å denotes height of pinholes present on top of domains. Same was estimated separately (56 Å) from a smaller AFM scan (200 × 200 nm²) of these portions only (not shown). Peak values obtained are close to those of 1ML CdSt distribution curve. This indeed shows that for 3ML CdSt film, deposition process is basically a continuation of 3D island formation. The 9ML counterpart (Fig. 7.3 (c)) is a continuation of the 3ML film with islands of various heights (62 Å, 168 Å , 203 Å , 223 Å and 243 Å). So, strictly speaking, there is *no change in surface morphology of CdSt with layer number*, just as in MnSt and CoSt. Thus it is seen that under this classification of morphological variation with layer number, MnSt, CoSt and CdSt falls in one group and ZnSt stands apart.

Evolution with Atomic Number of Metal

Comparing the AFM results of all four metal stearates, it is found that LB growth changes from 2D layering to commencement of 3D island formation with increase in atomic number (Z) of metal

Table 7.1: Calculated fractal dimensions D from AFM images

Film	D	Growth Mode
CdSt	1.41	V-W
ZnSt	1.49	S-K
CoSt	1.69	F-M
MnSt	1.62	F-M

from Mn to Cd. Moreover, island formation starts from first AML in CdSt, but only from the first SML in ZnSt (after deposition of a smooth AML) whereas, there are no island formation at all in MnSt or CoSt, where AMLs and SMLs show smooth morphology. Thus by studying their morphological evolution as regards commencement of island formation which depend upon atomic number of metal, the MSt films can be classified under the three growth modes observed in heteroepitaxy [81, 82], viz. Volmer–Weber (VW) or growth by 3D island formation as in CdSt, Frank-Van der Merwe (FM) or growth via 2D layering as in MnSt and CoSt and Stranski–Krastanov (SK) where island formation follows 2D layer-by-layer growth as in ZnSt.

Now, using the concept of fractal dimension, the morphologies of the different MSt multilayers have also been characterized. The fractal dimension is related to the scaling structure of a surface, and quantifies it. In thin film growth, the surface comprises islands, whose shapes and sizes govern the morphology of the film. The perimeter P of these islands, sectioned by a plane, relates to the surface area A of the islands as $P = \mu A^\alpha$ where μ is the proportionality factor between the perimeter and surface and $\alpha = D/2$, D being the fractal dimension. D values of the three films were extracted using WSxM software. These are given in Table 7.1. It is known that D is related to the growth mode, with D<1.5 denoting Volmer–Weber (VW) growth, D>1.5 denoting Frank van der Merwe (FM) growth and D~1.5 signifies Stranski–Krastanov (SK) growth. As seen in Table 7.1, D values of the 3ML films show that growth in CdSt films proceeds via VW mode, ZnSt shows SK growth and in CoSt/MnSt, growth occurs via 2D layering or FM mode.

Thus out of the above two classifications (I) and (II), the second appears to be of greater interest as this classification resembles those observed in general heteroepitaxial growth and hence finds a broader footing. However, since AFM, being a surface sensitive technique does not provide the actual values of the thicknesses of deposited layers, we have employed X-Ray reflectivity to obtain information about the layer thickness and their corresponding coverage, so as to confirm the aforementioned classification of growth modes.

7.1.2 X-Ray Reflectivity Results

X-Ray reflectivity curves (open circles) of MnSt (Fig. 7.4 (a)), CoSt (Fig. 7.4 (c)), ZnSt (Fig. 7.4 (e)) and CdSt (Fig. 7.4 (g)) were fitted (solid lines) using Parratt formalism [156], where each film is divided into layers of fixed thickness (d), average electron density (ρ) and interfacial roughness (σ), and used as fit parameters. Electron density profiles (EDPs) for films (ρ as a function of film thickness z, where z = 0 is air), were constructed from best fit values consistent with physical acceptability. The EDPs of MnSt (Fig. 7.4 (b)), CoSt (Fig. 7.4 (d)), ZnSt (Fig. 7.4 (f)) and CdSt (Fig. 7.4(h)) are plotted as 3-D graphs. In each graph, the individual EDPs for 1ML, 3ML and 9ML films are plotted together to show the variation of ρ with layer number as well. It must be mentioned that fits were very sensitive to variations in headgroup electron density (~ 0.01 e/Å^3) and thickness (~ 0.1 Å), which provided required selectivity in extracting EDPs. However, as metal-containing headgroup layer in contact with substrate is convoluted with interfacial width (or roughness), it is almost impossible to distinguish these headgroups in most EDPs. As shown earlier, XRR data confirm assignments of growth modes by AFM analysis with more accurate information about depth profile. XRR profile and EDP of 1 ML MnSt, CoSt and ZnSt confirm their true monolayer features as the extracted film thicknesses were of typical monolayer values of 23 Å for MnSt, 22 Å for CoSt and 21 Å for ZnSt. Thicknesses estimated from fitted data of 3 ML films are 76 Å for MnSt, 75 Å for CoSt and 63 Å for ZnSt, which correspond

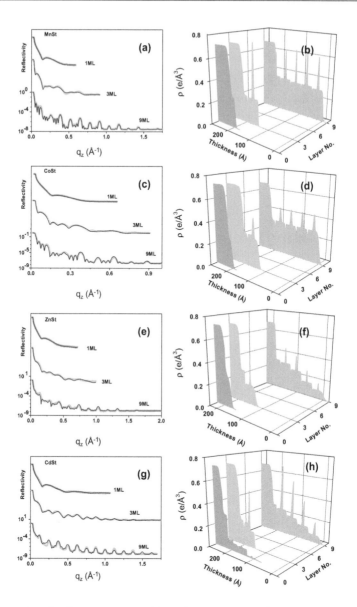

Figure 7.4: *Reflectivity profiles (open circles) with fitted curves (solid lines) of (a) MnSt; (c) CoSt; (e) ZnSt and (g) CdSt LB films. Curves for 1ML and 3ML have been upshifted for clarity. Electron Density Profiles of (b) MnSt; (d) CoSt; (f) ZnSt and (h) CdSt films plotted as 3D curves showing variation of average electron density ρ with film thickness and layer number.*

to typical trilayer thickness denoting formation of a SML on the AML. EDPs of both 3ML and 9ML films of MnSt and CoSt show no decrease in ρ with layer number suggesting uniform coverage and hence growth by 2D layering (FM type).

EDP of 3ML ZnSt shows a SML of low ρ on top of the AML. This shows that coverage of SML is much less compared to first AML, hinting at commencement of island formation after the first monolayer. In 9ML ZnSt film, coverage decreases with each SML, showing that island formation continues at least up to 9ML. ZnSt thus shows SK type growth.

An extra hump in the reflectivity profile of 1 ML CdSt indicates presence of islands forming on top of the AML. EDP shows existence of a comparatively low density SML on top of AML along with two more SMLs of negligible electron density, which were necessarily incorporated to fit the reflectivity profile. Thickness up to the first SML, obtained from fit, was 69 Å. Thickness of the film up to the first two SMLs was 114 Å whereas total film thickness was 158 Å. These thicknesses are in agreement with AFM data. XRR data of 3ML CdSt shows a trilayer feature of thickness 76 Å of hydrocarbon tail density (0.22 e/Å^3) much higher than that of the same of its 1ML counterpart (0.09 e/Å^3). Although presence of two SMLs on top of the trilayer feature is evident from the AFM morphology and distribution, owing to their negligible electron densities, they were not incorporated in the XRR fits. The 9ML film shows continuation of island formation with a decrease in ρ with layer number. AFM and XRR measurements of CdSt films confirm growth by 3D island formation, or more precisely, VW type growth. Thus, XRR results are throughout in agreement with those obtained from AFM measurements, as regards evolution of morphology of MSt multilayers.

7.2 Morphological Evolution and Metal–Carboxylate Headgroup

In this section, we have investigated the reason behind the dependence of morphology on atomic number. For this, we have investigated the structure of the metal-bearing headgroup using XRR and FTIR. For each metal stearate, the number (n) of metal atoms per headgroup viz. $n = 1$ (Mn/Co), $n = 0.75$ (Zn) and $n = 0.5$ (Cd) was calculated. Parameter n is found to decide headgroup coordination such that $n = 1.0$ corresponds to bidentate and $n = 0.5$ corresponds to unidentate coordination; intermediate value in Zn corresponding to a mixture of both. our results clearly indicated association of bidentate coordination of metal–carboxylate headgroup to layer-by-layer growth as observed in MnSt, CoSt and partially in ZnSt. Crossover to island-like growth, as observed in CdSt and ZnSt, is associated with presence of unidentate coordination in headgroup. Dependence of growth mode on headgroup structure is explained by the fact that in bidentate headgroups, in-plane dipole moment being zero, intermolecular forces between adjacent molecules are absent and hence growth proceeds via layering. On the other hand, in unidentate headgroups, existence of non-zero in-plane dipole moment results in development of weak in-plane intermolecular forces between adjacent molecules causing in-plane clustering leading to island-like growth.

7.2.1   Role of Metal–Carboxylate Ratio

In order to find the reason behind the dependence of film morphology on atomic number of metal, the metal-bearing headgroup should be studied in details. From XRR data, we have calculated the number of electrons per headgroup. Calculations on metal-bearing headgroup were carried out with fitted XRR data, assuming uniform molecular cross-sectional area A of 20 Å^2 [47]. Number of electrons N_H (N_T) in Head (Tail) for each layer of a film was calculated as $\rho_H \times A \times d_H$ ($\rho_T \times A \times d_T$), where suffix H (T) refer to head (tail). To eliminate effect of porosity, the ratio $R_{Exp} = N_H / N_T$ was computed for each layer. Values of R_{Exp} for all films are given in Table 7.2. In addition, theoretical R-values (R_{Th}) for all films were calculated. Taking into account the symmetry of SMLs with respect to head and tails, and the lack of it in AML, an appropriate model of a molecule is constructed. With this model, R_{Th} was calculated for all films, varying number of metal ions in

Table 7.2: Results of XRR analysis

M-St	AML			Successive SMLs			
	Film	R_{Expt}	R_{Th}	n	Film	R_{Expt}	R_{Th}	n
CdSt	1ML	0.36	0.34	0.5	3ML (1)	0.66	0.56	0.50
	3ML	0.40	0.34	0.5	9ML (1)	0.78	0.67	0.50
	9ML	0.44	0.34	0.5	9ML (2)	0.84	0.67	0.50
					9ML (3)	0.81	0.67	0.50
					9ML (4)	0.79	0.67	0.50
ZnSt	1ML	0.26	0.19	1.0	3ML (1)	0.25	0.28	0.75
	3ML	0.38	0.32	1.0	9ML (1)	0.25	0.28	0.75
	9ML	0.18	0.19	1.0	9ML (2)	0.27	0.28	0.75
					9ML (3)	0.30	0.28	0.75
					9ML (4)	0.30	0.28	0.75
CoSt	1ML	0.30	0.30	1.0	3ML (1)	0.52	0.60	1.00
	3ML	0.26	0.30	1.0	9ML (1)	0.46	0.60	1.00
	9ML	0.24	0.30	1.0	9ML (2)	0.48	0.60	1.00
					9ML (3)	0.48	0.60	1.00
					9ML (4)	0.48	0.60	1.00
MnSt	1ML	0.25	0.28	1.0	3ML (1)	0.96	0.69	1.00
	3ML	0.38	0.28	1.0	9ML (1)	0.23	0.32	1.00
	9ML	0.11	0.16	1.0	9ML (2)	0.23	0.32	1.00
					9ML (3)	0.25	0.32	1.00
					9ML (4)	0.24	0.32	1.00

model headgroups such that R_{Th} values matched those of R_{Exp} ones. Total number of electrons in model headgroup, for which R_{Th} was nearly equal to R_{Exp}, was thus estimated. From this, number of metal ions (n) per headgroup or carboxylate (COO) group is calculated for each layer in all deposited films. Results are given in Table 7.2. In CoSt and MnSt, $n = 1.0$ for both AML and SML headgroups. However in ZnSt, $n = 1.0$ for AML and $n = 0.75$ for SML headgroup. In CdSt, R values matched well for $n = 0.5$ for both AML and SML. It is noted that for each layer, in all films, $n = 1.0$ corresponds to growth by layering, and for values less than 1.0, there is signature of island formation. Thus we found that in deposition of each LB layer, for all MSt films, the morphology has a strong dependence on molecular configuration. Next, we try to find the cause behind such dependence, by elucidating the headgroup structure.

The number of metal ions (n) per COO group calculated from XRR gives insight into metal–carboxylate coordinations in films. There are three types of metal–carboxylate coordinations, viz. unidentate coordination, bidentate bridging coordination and bidentate chelate coordination [161]. Unidentate coordination results in contribution of half metal per COO group, whereas bidentate (both bridge and chelate) have one metal per COO group. The values of n, as obtained from XRR suggest that CoSt and MnSt prefer bidentate coordination in all layers, whereas CdSt prefers unidentate coordination. However, the situation is somewhat mid-way in ZnSt. Whereas, in its AML, ZnSt would prefer bidentate coordination, metal ion per headgroup contributions in ZnSt SMLs probably suggest a mixture of coordinations in these layers.

7.2.2   Fourier Transform Infrared Spectroscopy Results

In order to directly look at metal–carboxylate coordinations, FTIR spectroscopy in ATR mode was carried out. FTIR spectra for (I) MnSt and (II) ZnSt are shown in Fig. 7.5(a). ATR measurements for CoSt and CdSt samples have been discussed in Chapter 5. To find coordination of metal ion with carboxylate group, difference in symmetric (ν_s) and asymmetric (ν_a) COO stretching frequencies (Δ) are measured [161]. These frequencies are separately shown for all four metal stearates (Fig. 7.5 (b): (I) MnSt; (II) CoSt; (III) ZnSt and (IV) CdSt).

For both MnSt (symmetric stretch at 1397 cm^{-1}, asymmetric stretch at 1541 cm^{-1} and $\Delta = 144$ cm^{-1}) and CoSt (symmetric stretch at 1397 cm^{-1}, asymmetric stretch at 1540 cm^{-1} and $\Delta = 143$ cm^{-1} coordination is bidentate bridged, consistent with $n = 1$. On the other hand, ZnSt shows presence of two asymmetric stretches at 1538 cm^{-1} and 1595 cm^{-1} and symmetric stretch at 1397 cm^{-1} giving $\Delta_1 = 141$ cm^{-1} and $\Delta_2 = 198$ cm^{-1}, thereby confirming existence of both bidentate bridged and unidentate coordinations. Equal intensity of peaks (Fig. 7.5 (a)- (II)) indicate that they are probably present in the same ratio, which gives an av-

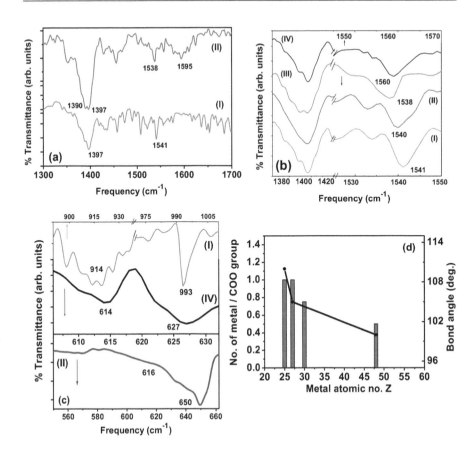

Figure 7.5: *(a) FTIR spectra of (I) MnSt and (II) ZnSt; portions of spectra showing (b) COO and (c) O-M-O (M=Mn,Co,Zn,Cd) stretching frequencies for (I) MnSt, (II) CoSt, (III) ZnSt and (IV) CdSt films; (d) variation of number of electrons (n) per COO group (shown as histogram) and metal carboxylate bond angle (plotted as points joined by line) with atomic number Z of metal for MSt films.*

erage contribution of three-quarter metal per carboxylate ion i.e., $n = 0.75$. For CdSt, symmetric and asymmetric stretch bands are at 1397 cm^{-1} and 1560 cm^{-1} respectively, giving $\Delta = 163$ cm^{-1}, that corresponds to unidentate coordination in which there are two carboxylate groups per metal ion.

Thus we find that the observed systematic variation of film

morphology on atomic number of metal stems from the association of bidentate coordination to growth by layering and unidentate coordination to island formation. Moreover, formation of unidentate coordination in headgroups can commence at any point of growth, whence island formation starts and continues to increase with layer number; the crossover from bidentate to unidentate being irreversible at least up to nine MLs. Again, a trend is observed in commencement of islandlike growth in moving towards metal of higher atomic number, as this leads to lower n.

Again, from AFM, XRR and FTIR results, it is clear that change in morphology with coordination can commence at any layer and so it can be postulated that coordination affects growth in the in-plane direction. The questions that come in mind are (1) How does headgroup coordination affect in-plane growth, and (2) Why does coordination change with change in atomic number of the transition metal? In order to address the first question we have tried to construct suitable models for the bidentate and unidentate headgroups, for both AML and SML from a knowledge of their structure and bonding obtained from XRR and FTIR, and tried to calculate the dipole moment of individual headgroups in order to estimate the in-plane long range forces between molecules. The answer to the second question comes as a logical follow up. These are discussed in the next subsections one by one.

At this point, we must mention that oxygen-metal-oxygen bond angle has been extracted from FTIR data based on metal-oxygen symmetric (ν_s) and asymmetric (ν_a) stretching frequencies. Details of such calculation is given in Chapter 6. Assigned stretching frequencies for (I) MnSt, (II) CoSt and (IV) CdSt are shown in Figure 4 (c). For MnSt ($\nu_a = 993$ cm^{-1} and $\nu_s = 914$ cm^{-1}), O-Mn-O bond angle is 110°; for CoSt ($\nu_a = 650$ cm^{-1} and $\nu_s = 616$ cm^{-1}), O-Co-O bond angle is 105° and for CdSt ($\nu_a = 627$ cm^{-1} and $\nu_s = 614$ cm^{-1}), O-Cd-O bond angle is 100°, i.e., O-M-O angle is found to decrease with increase in metal atomic number. However as ZnSt does not have a unique coordination, conformation angle is also not unique and hence bond angle O-Zn-O is not determined. A plot of this bond angle, along with variation of n with atomic number Z of corresponding metal is shown in Fig. 7.5

(d). Change in this angle may indicate change in hybridization state for these films [181] but a comparison with atomic structures of these transition-metals leads us to conclude that there is no straightforward relation between electronic structure of these metal atoms (in particular number of d-electrons) and either multilayer morphology or headgroup coordination. We have, on one hand, Mn and Co, having different number of d-electrons, giving rise to identical coordination and LB morphology, and on the other, Zn and Cd, having same number of d-electrons, leading to different coordinations and morphologies. Nevertheless, an estimate of the O-M-O bond angles are necessary for determination of the dipole moments of the headgroups, as discussed in the next subsection.

7.2.3 Role of Headgroup Dipole Moment

As seen above, in all four MSt films, the headgroups either have bidentate or unidentate coordinations. Moreover, structure of SML headgroups (Fig. 7.6(a)) are inherently different from their AML counterparts (Fig. 7.6 (b)). For SML headgroups, with a cross-sectional area of 20 $Å^2$ having non-linear O-M-O bond angle, possible structures for bidentate bridged (I) and unidentate (II) headgroups are constructed. From NEXAFS studies of MSt films, we have shown in Chapter 6 that in bidentate (unidentate) headgroups the carboxylate group has equal (unequal) CO bond strength, such that both COO groups are symmetric (asymmetric) in nature. In case of bidentate headgroups, two pairs of symmetric CO bonds leads to contribution of zero dipole moment coming from the two carboxylate groups in the SML headgroup (Fig. 7.6 (a) (I)). Also, the two O-M-O bonds being symmetric in nature, give a resultant of zero dipole moment, causing net dipole moment of the bidentate SML headgroup to be zero.

On the other hand, in unidentate headgroups, asymmetric COO bonds causes a shifting of the negative charge of the carboxylate ion towards the oxygen atom to which the metal ion is bonded, causing development of unequal dipole moments (p_1 and p_2 (say)) of the CO bonds in a COO group, as shown in Fig. 7.6(a) (II).

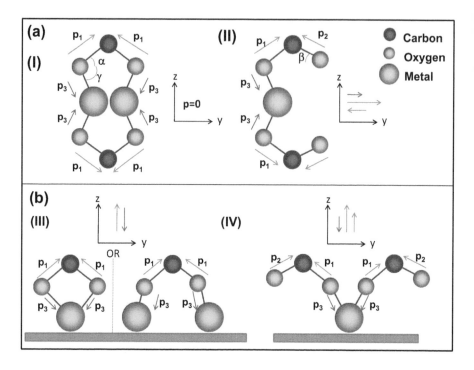

Figure 7.6: *Constructed models for (I),(III) bidentate and (II), (IV) unidentate coordinated headgroups in (a) SML and (b) AML of MSt films.*

Hence resultant dipole moment coming from the two COO groups is in the in-plane direction (y-axis) and has a magnitude $p_{COO} = 2 (p_1 \cos \alpha - p_2 \cos \beta)$, where α (β) is the angle made with the y-axis by C-O bond, bonded (not bonded) to the metal ion. Again, since there is only one O-M-O linkage, its net dipole moment is also along y-axis and has a magnitude $p_M = 2 p_3 \cos \gamma$, where p_3 is the dipole moment of individual M-O bond and γ is the angle made by M-O bond with the in-plane direction. Thus, in uniden-tate headgroups, net dipole moment is given by $p_{UC} = p_{COO} + p_M$ and acts totally in-plane.

Headgroup structures (both bidentate (III) and unidentate (IV)) in AML (Fig. 7.6 (b)) are different from those in SMLs. Here, for the bidentate configuration (both bridge and chelate), net dipole moment is given by $p_B = 2(p_3 Sin\gamma - p_1 Sin\alpha)$, out-of-plane and towards the substrate. For AML-unidentate head-group, net dipole moment has both in-plane as well as out-of plane components. While the out-of-plane component has a magnitude $p_{out} = 2(p_3 Sin\gamma - p_1 Sin\alpha - p_2 Sin\beta)$ towards substrate, the in-plane component p_{in} is vector sum of in-plane components of dipole moments of the two COO groups (since they may not be coplaner) making the latter non-zero.

From a knowledge of the O-M-O bond angles, estimated from FTIR studies, and from the fact that both O-M bonds are symmetric, the angle γ is determined. Values of γ for CdSt, CoSt and MnSt are 50 °, 52.5 ° and 55 ° respectively. Taking p_3 as the difference in electronegativity values of oxygen and metal, values of p_M are calculated to be 2.25 D for CdSt, 1.90 D for CoSt and 2.17 D for MnSt. Since the O-M-O linkage is not known, value of p_M for ZnSt could not be calculated. Again, since the values of the angles α and β cannot be obtained from these studies, it is not possible to estimate p_{COO} and hence p_{UC} in this case. Even a theoretical estimate of p_{UC} cannot be given, as the angle between the O-M-O linkage and the O-C-O linkage (which lie in different planes), is also not known. This has been discussed in Chapter 6.

Nevertheless, it is found that in both AML and SML head-groups, bidentate configuration has no in-plane component whereas unidentate configuration has a non-zero counterpart, the latter being quite appreciable as suggested by the high value of p_M for CdSt. It is this non-zero in-plane component of dipole moment arising in individual headgroups that give rise to in-plane inter-molecular dipolar interactions among adjacent headgroups, dictating their in-plane molecular growth. More specifically, due to dipole-dipole interactions between adjacent headgroups, unidentate system has stronger intermolecular attraction than molecule-substrate attraction, leading to domain like growth with pinhole defects. In contrast, there exists no in-plane dipolar interactions in bidentate headgroups, such that these molecules do not prefer clustering over growth by 2D layering. Moreover, in both type of AML headgroups, there exists a non-zero component of dipole moment perpendicular to the substrate pointing towards it, which probably leads to adhesion of film to the substrate. However, in both type of SML headgroups, there exists no component of dipole moment in the out of plane direction, stating that dipolar interactions do not play a part in formation of LB multilayers; the latter being solely governed by supramolecular tail-tail interactions of van der Waals and hyperconjugative origins as we have suggested previously (in Chapter 6).

7.2.4   Role of Metal Atomic Number

It is worthwhile mentioning here that since in bidentate headgroups, the dipole moment of the system is in general lower compared to unidentate ones, the former is a much stabler system compared to the latter. It can thus be stated that transition metal ions in general prefer to coordinate with the carboxylate groups as bidentate bridge. In order to do so, it is necessary to accommodate an extra metal ion in the headgroup. With decrease in atomic number of the metal, the ionic radius decreases, such that only beyond a cut off value (near Zn in this case), two metal ions can be accommodated in a headgroup. It is probably due to this reason that headgroup coordination has a dependence on atomic number, which answers the second question.

It must be noted that this dependence of n on Z is basically limited to relatively lower pH of the aqueous subphase on whose surface the Langmuir monolayers of stearic acid bond with the M-ions dissolved in subphase. The values of n can be manipulated with higher pH [194] and a bidentate-bridged headgroup with consequent pinhole-free growth can be achieved even for Cd-ions, showing that crucial dependence of headgroup coordination and LB growth mode on n.

As pointed out above, absence of headgroup-headgroup interactions in MnSt/CoSt SMLs, along with FM type "fully wetting" films obtained from them, point to a "liquid-like" character of these layers, in particular, where molecules are in symmetric configuration. It is to be noted that for preformed three-tailed amphiphiles such as ferric stearate arranged in similar fashion [195] or in collapse of MnSt/CoSt monolayers on water surface [63] exactly similar liquid-like behaviour has been observed for symmetric configuration. Again, same island-like growth was observed for CdSt SMLs [63] and here we have explained this growth as due to the in-plane intermolecular attraction. However, in-plane height correlations for CdSt show a "liquid-like" behavior at large length scales [57] and a self-affine behavior at short length scales [73]. Hence, similar studies need to be carried on MnSt, CoSt and ZnSt for comparison and those are underway.

7.3 On Morphological Evolution

Morphological and structural studies clearly indicate that for at least four transition metal stearate Langmuir–Blodgett (LB) films, viz. MnSt, CoSt, ZnSt and CdSt, bidentate coordination of metal–carboxylate headgroup gives rise to layer-by-layer growth as observed in Mn, Co and partially in Zn. Crossover to island-like growth, as observed in Cd and Zn, is due to presence of unidentate coordination in headgroup. Morphological evolution of each MSt film is in accordance with its previously assigned growth mode, with MnSt/CoSt showing FM type growth, ZnSt showing SK type growth and CdSt showing VW type growth. Growth mode is found to vary with number (n) of metal atoms per headgroup, viz. $n = 1$ (MnSt/CoSt), $n = 0.75$ (ZnSt) and $n = 0.5$ (CdSt), which in turn decide headgroup coordination such that $n = 1.0$ corresponds to bidentate and $n = 0.5$ corresponds to unidentate coordination; intermediate value in Zn corresponding to a mixture of these coordinations. These results are summarized in Fig. 7.7. Moreover, it is found that in bidentate headgroups, in-plane dipole moment was zero, which causes intermolecular forces between adjacent molecules to be absent and hence allowing growth to proceed via layering. On the other hand, in unidentate headgroups, existence of non-zero in-plane dipole moment resulted in development of weak in-plane intermolecular forces between adjacent molecules causing island-like growth. Again, while in AML, dipolar forces perpendicular to substrate is found to exist in both headgroup coordinations, probably aiding metal-substrate adhesion, the same

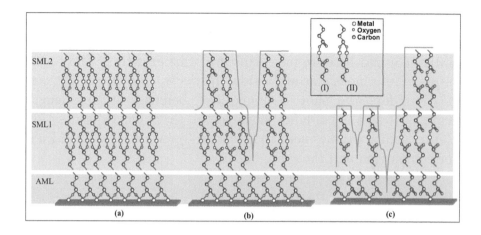

Figure 7.7: *Cartoon showing different growth modes and coordinations in deposited films: (a) Frank van der Merwe type in MnSt/CoSt, (b) Stranski–Krastanov in ZnSt and (c) Volmer–Weber in CdSt. The coordinations are separately shown (inset) as (I) unidentate and (II) bidentate bridged. The film surface morphology is shown by the bold line.*

is found to be absent in all SMLs, suggesting that dipolar interactions arising in the headgroup do not affect LB multilayer formation. Gradual decrease in n, in going from Mn to Cd, suggested that crossover from layering to island formation is not an abrupt process but rather systematic with change in atomic number of metal. Moreover, it is argued that dependence on atomic number probably stems from the dependence of coordination on ionic radius of metal, the latter decreasing with decrease in atomic number.

Chapter 8

"Liquid" and "Solid" Films

Morphological evolution of metal stearates show distinct growth modes that stem from the fact that amphiphiles of this type have either unidentate or bidentate coordination in their headgroup. More specifically, absence of in-plane dipolar forces in MnSt/CoSt headgroups, along with FM type "fully wetting" films obtained from them, point to a "liquid-like" character of these layers, in particular, where molecules are in symmetric configuration. In sharp contrast to this, the CdSt and ZnSt films exhibit VW type "island" growth along with presence of in-plane dipolar forces in their headgroup. In the previous chapters, we have shown how the coordination and hence the geometry of the metal–carboxylate headgroup decides the presence or absence of long-range intermolecular forces through the presence or absence of dipole moment, and have hinted that it may control the liquid-like or solid-like behavior of the corresponding multilayers. In this chapter, we will compare the height correlations in monolayers and multilayers of amphiphilic salts of two different divalent metals, Cd and Co, deposited on amorphous silicon substrate to bring out their essential differences as "solid" or "liquid"-like, respectively.

We have determined the height difference correlation functions from topographic images of CdSt and CoSt LB films from tapping mode AFM. Results show distinct difference in their in-plane morphologies. CdSt films, with huge number of in-plane "pinhole" defects, follow self-affine behavior whereas CoSt films, which are almost void of such in-plane defects, show deviation from self-affinity especially at small length scales, suggesting liquid-like behavior, imparting flexibility to the system, in-plane. Phase images of CoSt obtained from tapping mode AFM show gentle undulations or hemisphere-like features in contrast to its smooth topography, unlike the CdSt system where both height and phase images show self-affine domains. NEXAFS spectoscopic studies indicate no preferred in-plane orientation of the headgroup in CoSt films. The undulating features in CoSt is explained by invoking a radially symmetric orientational distribution in the tilt of adjacent hydrocarbon tails, causing a small in-plane density variation which shows up in the phase image. These orientational disorders in adjacent tails probably allows "filling up" of in-plane defects which gives rise to its excellent in-plane coverage and hence a "liquid" like behavior in CoSt.

8.1   Metal Organics and "Liquid"-Like Behavior

8.1.1   Thin Film Preparation and Measurements

One and three MLs each of CdSt and CoSt were deposited on hydrophilic silicon (100) substrate by the LB technique described in Chapter 2. 0.5 mM chloride solutions of Cd^{2+} and Co^{2+}) ions were used. Subphase pH was maintained at 6.0 by adding sodium bicarbonate. stearic acid solution (0.5 mg/ml) was spread to form the monolayer. Films were deposited at a monolayer pressure of 30 mN/m at 19°C at a dipping speed of 3 mm/min, the first layer being deposited by an upstroke of the substrate from water to air through the interface. Drying time after first stroke was 10 min. Films were checked for reproducibility. AFM was performed in tapping mode. Scans were performed over several regions of the films for different scan areas. NEXAFS spectroscopy [147] at oxygen K-edge ($\sim$ 530 eV) was carried out at normal and grazing incidence using synchrotron radiation. The spectra were recorded by total electron yield (TEY). They were normalized to an I_0 spectrum previously measured for clean copper to account for oxygen in mesh. Monochromator slits (50 μm) were set to provide an energy resolution of 0.1 eV.

8.2 "Solid" and "Liquid"-Like Langmuir–Blodgett Films

8.2.1 Height Correlations on Film Surface

AFM topographic images of 1ML (scan size (S): 1×1 μm^2) and 3ML (S: 5×5 μm^2) films of CdSt and CoSt are shown in Fig. 8.1. As seen from the topographic images the CdSt films, both 1ML (Fig. 8.1(a)) and 3ML (Fig. 8.1(c)), show presence of a

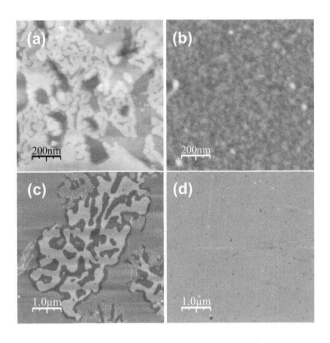

Figure 8.1: *AFM topographic images of (a), (b) 1ML (1 × 1 μ m^2) and (c), (d) 3ML (5 × 5 μ m^2) films of (a), (c) CdSt and (b), (d) CoSt.*

Table 8.1: Conversion factor

Scan size (μm)	m	r(nm)
0.2	1	0.78
0.5	1	1.95
1.0	1	3.91
2.0	1	7.81
5.0	1	19.53
10.0	1	39.06

large number of in-plane "pinhole" defects, compared to the almost defect-free topography of CoSt (1ML (Fig. 8.1(b)) and 3ML (Fig. 8.1(d)). As discussed in Chapter 7, the CdSt and CoSt films resemble two different growth modes [81] viz., the Volmer–Weber (or island type) for the former and Frank van der Merwe (or wetting type) in case of the latter.

Although these growth modes relate to the out-of-plane growth of the films, they are dependent on the formation of "pinhole" defects in this type of LB films, the latter also deciding the in-plane growth or surface morphology. In order to study the difference in surface morphology of the two types of metal-bearing films, a height-difference correlation function $g(r)$ (see Eqn. 4.1) was calculated from the AFM topographic images for various scan sizes for 1ML and 3ML films of CdSt and CoSt. To plot curves of different scan sizes in a single graph, instead of $g(r)$ vs r, we have plotted curves of $g(m)$ vs m, where m stands for the number of pixels in the AFM image. For a particular scan size, one pixel corresponds to a specific value of the length r in nm, the conversion factor being given in Table 8.1. The $g(m)$ vs m curves are plotted for CdSt (Fig. 8.2(a) and (c) for 1ML and 3ML films respectively) and CoSt (8.2 (b) and (d) for 1ML and 3ML films respectively) corresponding to different scan sizes (viz. (I) 0.5μm^2, (II) 1.0μm^2 and (III) 2.0μm^2). For the 3ML films, the same are also plotted for 5.0μm^2 and 10.0μm^2 scan sizes (inset).

The curves for CdSt and CoSt were fit with the correlation function for self-affine roughness (Eqn. 4.2) and the parameters of fit are given in Table 8.2. As seen from the figures, curves for 1ML and 3ML CdSt (Fig. 8.2(a) and (c) respectively) show

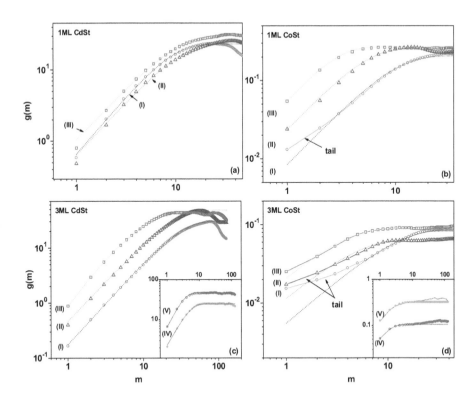

Figure 8.2: *Height-difference correlation function of (a), (b) 1ML and (c), (d) 3ML films of (a), (c) CdSt and (b), (d) CoSt for (I) 500 × 500 nm²; (II) 1 × 1 μ m²; (III) 2 × 2 μ m²; (IV) 5 × 5 μ m² and (V) 10 × 10 μ m² scan size. The symbols represent the data estimated from AFM topographic images, while the dotted lines are the fit corresponding to Equation 4.2.*

excellent fits to Eqn. 4.2 and hence they strictly follow self-affine behavior for all length scales. It should be mentioned here that this is consistent with previous results regarding height correlations obtained in CdSt films obtained from AFM measurements [57,73].

However, the height correlations obtained from diffuse x-ray scattering, which covers a much larger length scale, deviate from self-affinity even for CdSt films and show a liquid-like behavior [73, 196, 197]. The 1ML and 3ML CoSt curves, on the other hand, deviate from Eqn. 4.2 (Fig. 8.2(b) and (d) respectively) at small r and are fit separately (discussed below). Nevertheless, CoSt

Table 8.2: Parameters of fit with equation 4.2

Film	Scan size (μm)	CdSt			CoSt		
		σ(Å)	ξ(nm)	H	σ(Å)	ξ(nm)	H
1ML	0.5	34.0	4.4	0.83	0.32	4.6	0.72
	1.0	35.0	2.8	0.76	0.36	1.2	0.75
	2.0	39.3	1.3	0.74	0.37	0.3	0.84
3ML	0.5	39.0	15.9	0.75	2.1	5.6	0.57
	1.0	49.0	6.4	0.72	1.8	1.1	0.57
	2.0	47.3	1.4	0.78	2.1	0.4	0.60
	5.0	35.0	0.3	0.78	2.3	0.1	0.60
	10.0	48.0	0.1	0.85	4.0	0.1	0.57

films obey self-affine nature at large length scales and hence, a comparison of the fitting parameters of Eqn. 4.2 for CdSt and CoSt are discussed. As seen from Table 8.2, the r.m.s. roughness σ is considerably more for CdSt compared to CoSt for both the monolayer and trilayer films for all scan sizes, thereby indicating that the CoSt surface has a smooth morphology with negligible height variation. This is consistent with our previously reported results regarding the excellent out-of-plane growth of these type of films. Moreover, as seen in Table 8.2, in CoSt, the correlation length ξ falls faster to zero than in CdSt, for all scan sizes, both for the 1ML and 3ML films.

The Hurst exponent H is found to considerably decrease from ~ 0.8 to ~ 0.6 in going from the 1ML CoSt to the 3ML CoSt film in contrast to the constant value ~ 0.8 in case of CdSt films. H is related to the fractal dimension D as $D = d - H$, where d is the dimensionality of the system [197]. For thin films, where the height variations are orders of magnitude less than the in-plane surface coverage, $d \sim 2$. In such case, for a V-W surface, $D < 1.5$ so that $H > 0.5$ and for an F-M surface, $D > 1.5$, so that $H < 0.5$. The H-values obtained in our case (Table 8.2), consistent with our results discussed in Chapter 7, not only indicate this trend but also show that in case of CoSt, in going from the 1ML to the 3ML film, the effect of substrate is reduced and that 2D layering becomes more pronounced.

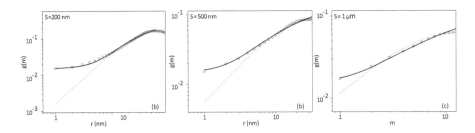

Figure 8.3: *Height-difference correlation function of 3ML CoSt films for (I) 200 × 200 nm²; (II) 500 × 500 nm² and (III) 1 × 1 μ m² scan size. The symbols represent the data estimated from AFM topographic images, while the dotted lines are the fit corresponding to Equation 4.5.*

As mentioned earlier, the CoSt curves do not strictly obey self-affinity. Deviation from self-affinity is evident for S= 500nm for the 1ML film. The effect is somewhat enhanced for the CoSt trilayer, at lower scan sizes viz. $0.2\mu m^2$, $0.5\mu m^2$ and $1.0\mu m^2$, all of which show presence of a "tail" or non-linearity (Fig. 8.3) a clear deviation from self-affine behavior. The "tails" are separately fitted by the logarithmic function (Eqn. 4.5) discussed in Chapter 4. The values of these parameters (Table 8.3) show that with increase in scan size, the r^2 term becomes more dominant, while the parameter b_p remains more or less constant. Although it is not possible to give a detailed description of this function at this point, we have tried to relate this dependence to the observed density variation, discussed later in this section. Nevertheless, this logarithmic dependence [73] as well as the absence of long-range intermolecular forces in CoSt films, as discussed in Chapter 7, suggest a probable liquid-like behavior in CoSt films in addition to self-affinity (at larger length scales).

Table 8.3: Parameters of fit with equation 4.5

Scan size (μm)	$a_p (nm^{-2})$	$b_p (nm^{-1})$	c_p	$d_p (nm^2)$
0.2	0.009	-0.001	1.343	0.052
0.5	0.150	0.050	2.250	0.018
1.0	0.450	0.050	2.500	0.016

It must be mentioned here that although the deviation from self-affinity is observed for small scan sizes, our results do not suggest that the effect is completely lost at large scan sizes but rather is masked due to increase in the number of "pinholes" (although still very few compared to its CdSt counterpart). This deviation from self-affinity especially at small length scales is significant and indicates the difference in the microscopic organization of the metal-bearing amphiphiles. Again, that the effect is found to be enhanced for the 3ML films from the 1ML film is consistent with the idea that the "liquid-like" behavior is an inherent characteristic of the amphiphiles and is suppressed to an extent by interaction with hydrophillic substrate.

8.2.2 Density Variation on Multilayer Surface

The difference between the films of these two metal-bearing amphiphile becomes more stark if the topographic and phase images of 3ML films of CdSt (Fig. 8.4 (a) and (b) respectively) and CoSt (Fig. 8.4 (c) and (d) respectively) obtained from the tapping mode of AFM, are compared. CoSt phase image shows the presence of prominent "hemisphere-shaped" features all over the film covered surface, which are near absent in its corresponding topographic image. Against this, in CdSt at regions of full coverage, the phase image does not show such prominent "hemispheres" and both topographic and phase images show similar, "island-like" features. This suggests the presence of a gentle undulation over the film covered surface in CoSt, which clearly does not arise due to interfacial roughness between two adjacent layers along z. The small r.m.s. roughness and also our previous results, as discussed in Chapters 5 and 7, suggest that successive LB multilayers of this type have smooth interfaces and hence do not affect the top roughness of the film covered surface. Thus a likely origin of such undulations in CoSt is in-plane density variation in the hydrocarbon chains. we propose that this density variation (observed in the nm length scale) is the result of a radially symmetric orientational disorder in the tilt of the hydrocarbon tails in CoSt. Although an individual

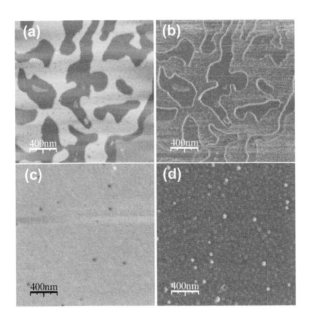

Figure 8.4: *AFM images (2 × 2 μ m²) of 3ML films of (a), (b) CdSt and (c), (d): CoSt; where (a), (c) and (b), (d) represent topographic and phase images respectively.*

molecular tilt does not give rise to significant density change at that length scale, the same occurring in a radially symmetric manner and over considerable area ($\sim$ nm) may cause an appreciable variation in density.

This can be explained by considering two cylinders of volume V, each containing N molecules, having radius R_0 and height S_0 (height of untilted molecular layer), placed adjacent to each other. The density then corresponds to density of untilted molecules given by $\rho = N/V$. Next we consider that the molecules in the cylinders are tilted by angle θ in a radially symmetric fashion such that the cylinders take the shape of two frusta of radius R such that $R > R_0$ for the first and $R < R_0$ for the second (Fig. 8.5), h being the height of both frusta. The change in density $\Delta\rho$ is then given by,

$$\Delta\rho = \frac{18a}{\cos\theta[9 - 3a^2 + a^4]}\rho \qquad (8.1)$$

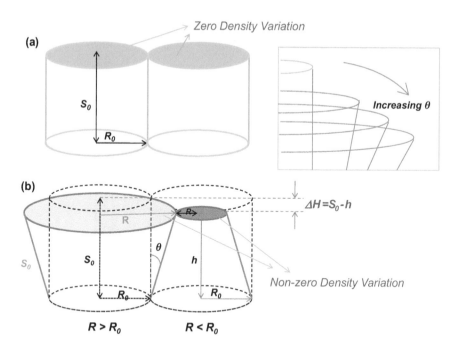

Figure 8.5: *Model for explaining height and density variation in CoSt films. (a) Two cylinders of radius R_0 and density ρ representing uniform distribution of untilted molecules; (b) Two frustums of unequal radii R formed from the cylinders by molecular tilt showing density variation; height variation within the frustum of radius $R > R_0$ due to formation of concentric frustums (inset).*

where $a = (S_0/R)\sin\theta$. Assuming $S_0 = 5$ nm to be the typical SML height and $R = 25$ nm to be the radius of hemispheres observed in AFM phase image, we plotted the variation of the percentage change in density with tilt angle θ (Fig. 8.6). It is found that even for small amount of tilt ($\sim 20°$), the percentage variation is appreciable ($\sim 20\%$).

It must be noted here that there is no relative height difference between the two cylinders or the two frusta, showing that the radially symmetric tilt of molecules by an angle θ do not contribute to any height variation. However, within any particular frustum whose radius $R > R_0$, there are a large number of

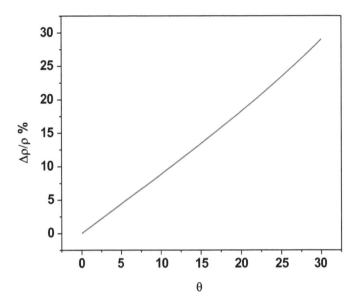

Figure 8.6: *Graph of percentage density variation with molecular tilt.*

concentric frusta with tilt angles varying from 0 to θ_{max}. This tilt variation leads to a small height variation given by $\Delta h = S(1 - cos\theta)$(Fig. 8.5(inset)), which becomes prominent only for small scan sizes and gets masked when "pinholes" become dominant in a system. We propose that in case of CoSt, for small scan size (where pinhole defects are insignificant), the AFM tip senses both the height and density variation arising due to molecular tilt. In the height image, they manifest themselves as "tails" and in the phase image they emerge as the "hemisphere-like" features. We find that $1/b_p$, where b_p is the coefficient of the linear term of Eqn. 4.5, has a value ~ 20 nm, which is of the order of the average radius R of the hemispheres. The first term of Eqn. 4.5 probably gives the total number of such correlated frusta with $R > R_0$, and hence this parameter a_p increases with an increase in scan size. As we shall see in the next section, our previous results (Chapters 6 and 7) and results of NEXAFS spectroscopic studies provide some validation for our model of frusta due to tilt variation.

8.2.3   Role of Molecular Orientation and Tilt

In Chapter 5 we have shown that multilayers of CdSt are untilted with respect to the surface normal whereas the same for CoSt are oriented at an angle of 9°. These values refer to spatial averages along the (x,y) direction. Hence they do not give information on the molecular orientation along this direction. In order to investigate the in-plane orientation of the molecules in CoSt films, NEXAFS spectra of 3ML CoSt was carried out in transmission mode at the O K-edge. Data was taken for normal and grazing (20° with horizontal) incidence. The polarization in the plane normal to the beam was at 0°, 30°, 45°, 60° and 90° for normal incidence, probing the in-plane orientation of the C-O bonds in the headgroups, while the electric field polarizations for grazing incidence were at 0°, 15°, 30°, 45°, 60°, 75°, and 90°, probing the bond orientations normal to the film plane. The series of measurements showed no variation of the main absorption peaks with change in polarization angle (Fig. 8.7), signifying no net preferred in-plane orientation of headgroups in the CoSt LB films. As the all-trans plane of the corresponding hydrocarbon chains are fixed i.e., rigidly attached to the headgroup w.r.t. the surface normal [47], the tails do not have an overall preferred orientation w.r.t. surface normal. This indicates that for the CoSt system, the hydrocarbon tails are tilted in a radially symmetric manner with an average tilt of 9° giving rise to "hemisphere-like" features in the phase images, with gradually increasing tilts from 0 to θ_{max}. The molecular tail tilt (Fig. 8.5), gives rise to both height and density variation, both becoming prominent only at small scan sizes, where presence of "pinholes" do not mask this subtle variation. Moreover, as is clear from Fig. 8.5, the height variation due to individual molecular tilt ($\sim$ for a tilt of 9°) is much less pronounced than the density variation, and is consistent with the fact that the variation is prominent in phase images than in height images.

We propose that this radially symmetric hydrocarbon tail tilt, although small ($\sim 0°$ *to* 18°) does give rise to a significant variation in density as is clearly brought out in the frusta model while the same model results in a weak variation in height of the frusta.

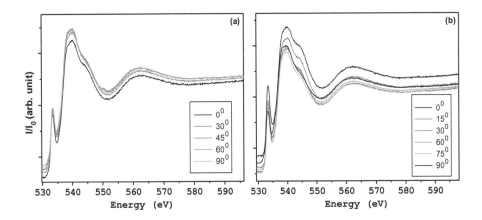

Figure 8.7: *O k-edge NEXAFS spectra of 3ML CoSt films at (a) normal and (b) grazing (20°) angles of incidence for different in-plane polarization angles.*

Thus the density variation caused by the variation in tail tilts are prominent in the AFM phase images but weak in the topographic images. This subtle height variation due to the variation in tilts also shows up as "tail"-like features in the height-height correlation function. It is this random orientation of tails that allows "space-filling" to occur along the in-plane direction in CoSt films reducing the "pinholes" in each layer as shown in the cartoon (Fig. 8.8). The connection between this reduction of the size of random desorption patches or pinholes with emergence of dominant liquid-like height correlations is consistent with x-ray diffuse scattering studies [196]. In this connection, the $1/b_p$ parameter in the height-height correlation function assumes a significance as the "wavelength" of the correlated frusta.

The next question that comes to mind is regarding the presence of these "liquid" and "solid" like features in the parent Langmuir monolayers of the metal-bearing amphiphiles. In order to investigate that, we have studied the metal-stearate monolayers at the air–water interface using Brewster Angle Microscopy (BAM), preliminary results of which have been discussed in Chapter 10 of this book.

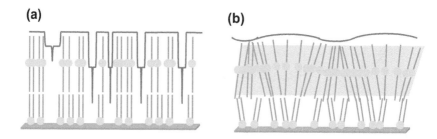

Figure 8.8: *Cartoon showing (a) untilted hydrocarbon tails in CdSt and (b) radially symmetric tilt orientation of hydrocarbon tails in CoSt films.*

To conclude, atomic force microscopic studies of CdSt and CoSt Langmuir–Blodgett films clearly bring out the difference in their in-plane structures. The CdSt films, with huge number of in-plane "pinhole" defects, exhibit self-affine behavior over a considerable scan range. On the other hand, the CoSt films, which are almost void of such in-plane defects, do not strictly follow the self-affine behavior. These films show deviation from self-affinity for small scan sizes and is pronounced for the trilayer film. The deviations probably suggest more of a liquid-like behavior and hence a much more flexible in-plane structure compared to CdSt films. The CoSt in-plane structure exhibits a smooth topography along with a gentle undulation, unlike the CdSt system, as seen from the AFM phase images. Such undulating topography in CoSt is probably achieved by the random orientation of the small tilts of adjacent hydrocarbon tails in a radially symmetric manner, as ascertained from NEXAFS spectoscopic studies. Thus in CoSt, tail orientations help fill up in-plane defects and give rise to excellent in-plane coverage.

Chapter 9

Metal–Organics and Nanocrystal Formation

LB films serve as potential low energy templates [49, 199–201]. They have an interesting property of changing from a low energy to a relatively high energy one through the absence or presence of dipoles carried by headgroups. This behavior can be utilized to manipulate the effect of these dipolar forces in tuning structures of metal–organics deposited on these templates. It is to be noted that in semiconductor heterostructures, the strain induced in the system can effectively tune the shape [202, 203] as well as band gap [204, 205] of nanocrystals. Similar effect may be observed in case of soft metal–organics, where template-overlayer forces induce nanocrystal formation.

In this chapter, we present results on a possible application of CdSt and CoSt trilayers as templates for self-assembling nanocrystals of similar metal-bearing amphiphiles. For this, zinc stearate (ZnSt) is deposited on the two metal–organic templates (MOTs) viz. Co-stearate (Co-T) and Cd-stearate (Cd-T) LB trilayers under identical conditions. Self-assembly of Zn-nanocrystals on both templates has been observed by AFM. Deposited crystals showed difference in structure on these two types of templates used. In order to elucidate the nanocrystal structure, diffraction measurements were performed in the vicinity of C K-edge using synchrotron radiation at BEAR beamline, Elettra. Results show that the ZnSt

Table 9.1: Characteristics of MSt templates

	Cd-T	Co-T
1. Growth	Island-like (V-W)	Layer by layer (F-M)
2. Headgroup	Unidentate	Bidentate
3. Dipole moment	Non-zero	Zero

nanocrystals have different structures when deposited on CdSt and CoSt templates, the former having non-close packed structure with superlattice ordering and the latter having a close packed structure without superlattice ordering.

The metal–organic templates (MOTs) in this case are three monolayers of CdSt and CoSt films, assembling into one SML sitting over one AML upon the substrate in each case. They have been characterized by AFM, XRR, NEXAFS and FTIR spectroscopic techniques as regards their morphology and structure. The results have been previously discussed in Chapter 7. However, for the sake of completion we have summarized the results once more in Table 9.1. The main differences between Cd-T and Co-T are that in the former, headgroup coordination is unidentate type whereas in the latter the coordination is bidentate type. Unidentate coordination in Cd-T leads to development of non-zero in-plane dipole moment and hence presence of in-plane intermolecular forces. This in turn causes island-like growth with a large number of "pinhole" defects in Cd-T. However, in Co-T, bidentate coordination in the headgroup causes in-plane dipole moment to be zero. This results in absence of intermolecular forces and in turn leads to space filling morphology void of "pinhole" defects. In spite of this difference, both CdSt and CoSt multilayers exhibit excellent periodicity along the out-of-plane direction, as discussed in Chapter 5, and hence may be used as suitable templates for deposition of similar metal–organics.

9.1 Nanocrystal Self-Assembly on Metal–Organic Templates

As seen from the previous sections, cadmium stearate and cobalt stearate fall under different catagories as regards their structure and morphology. Most importantly, in CdSt, long range inter-molecular forces are present along the in-plane direction whereas the same is absent in CoSt. The question that comes in mind is that whether these LB multilayers can be used as organic templates. Specifically, what effect will the presence or absence of in-termolecular forces in the template have on the deposited species? In order to understand the role of CdSt and CoSt as suitable templates, we have chosen to deposit zinc stearate on them by the LB technique. As seen in Chapter 7, ZnSt have mixed coordination of the headgroup in its SMLs when deposited on Si substrate. Since ZnSt has equal preference for both unidentate and bidentate coordinations, it can be assumed that it will possibly not be biased by the coordination of the template. By comparing the morphology and structure of deposited ZnSt on the two types of templates, we intend to find out the effect of the template structure on deposited ZnSt.

9.1.1 Preparation and Measurements

ZnSt was deposited on the Cd-T and Co-T templates using the same LB technique, in which divalent Zn ions were dissolved in the subphase and stearic acid was spread to form the Langmiur Monolayer (LM). The templates were vertically passed through the LM twice, starting from air. The films were characterized by FTIR to study bonding, AFM to study the morphology and x-ray diffraction in the vicinity of the C k-edge using synchrotron radiation to elucidate the crystalline structure.

9.1.2 Surface Morphology of Self-Assembled Zinc Stearate on Metal–Organic Templates

AFM topograghic images of ZnSt deposited on Cd-T and Co-T are shown in Fig. 9.1. In brief, ZnSt is found to self-assemble on both templates as grain-like entities with sharp edges having an average height of ~ 25 nm. However, as seen from the images, the morphology of ZnSt on Cd-T (Fig. 9.1(a)) is found to differ

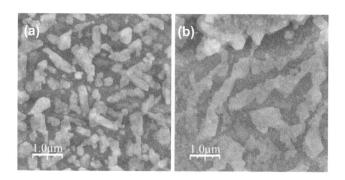

Figure 9.1: *AFM topographic images ($5 \times 5 \ \mu \ m^2$) of ZnSt nanocrystals on (a) Cd-T and (b) Co-T.*

from that on Co-T (Fig. 9.1 (b)). On Cd-T, ZnSt formed prominent grain-like features with edges having angles $\sim 120°$, $115°$ and $110°$. On the other hand, the ZnSt "grains" on Co-T showed less prominent "grain"-like structure with edges having angles $\sim 135°$, $120°$, $115°$, $105°$ and $75°$. Moreover, the grain size is smaller on Cd-T compared to that on Co-T. Above results shows that on Cd-T, ZnSt has a more regular structure than on Co-T and hence higher entropy. This shows that the same species (i.e., ZnSt), when deposited on different templates (Cd-T and Co-T) indeed show different morphology.

9.1.3 Headgroup Coordination and Bonding of Overlayer

Fourier transform infrared (FTIR) spectroscopic studies of the overlayer films and their templates were carried out to elucidate their headgroup structure. For this, the FTIR spectra of the ZnSt overlayer films and their templates have been studied by comparing the spectra with that of bulk stearic acid (Fig. 5.5). Both films show presence of two new peaks at $1730 \ \mathrm{cm}^{-1}$ and $1237 \ \mathrm{cm}^{-1}$, which are absent in the spectra of either the templates Cd-T/Co-T or bulk stearic acid. Although they lie close to the bulk acid peaks, they fall outside the range of the latter, and hence cannot be considered to be signature of undissociated stearic acid present in these films. Moreover, these peaks are sharper in case of films on Cd-T compared to those on Co-T, which is consistent with better "grain" formation on Cd-T, as observed from AFM studies. These are hence, attributed to the formation of "nano-grains" in the overlayer. We propose that they point to a characteristic acid headgroup structure corresponding to these "nano-grains," which co-exists with the unidentate and bidentate coordinations present at the metal–organic interfaces within the layers. This is discussed in details in the next subsection.

FTIR studies of the films also bring out the role of templates in deciding the headgroup coordination preference and hence the structure of the overlayer films. The metal-stearates deposited on Cd-T are strongly affected by the coordination of the template, whereas the ones on Co-T are not so. More specifically, ZnSt on Cd-T shows two peaks of unequal intensity at 1548 cm^{-1} and 1498 cm^{-1}, corresponding to two asymmetric stretches of the dissociated COO group, the former being more intense, with the symmetric stretch at 1383 cm^{-1}. ZnSt/Cd-T thus has two coordinations, unidentate ($\Delta = 165$) and bidentate ($\Delta = 115$), with preference towards unidentate coordination.

On the other hand, ZnSt/Co-T shows two peaks of more or less equal intensity at 1541 cm^{-1} and 1519 cm^{-1}, corresponding to two asymmetric stretches of the dissociated COO group, with the symmetric stretch at 1386 cm^{-1}. This indicates presence of two coordinations in ZnSt/Co-T, unidentate ($\Delta = 155$) and bidentate ($\Delta = 133$), with equal preference towards both. This shows that Co-T does not affect individual headgroup coordination preference unlike Cd-T, where the overlayer headgroup preference is strongly dictated by the template coordination type.

To elucidate the headgroup structure of the overlayer "nano-grains," we have compared the bond strength of the observed "nanocrystal" COO peaks with both dissociated and undissociated stearic acid peaks present in all films along with those of the templates and bulk stearic acid. For this, we have assumed the COO group to comprise of two diatomic systems C-O and C=O, the force constants were calculated for C-O (k_1) and C=O

(k_2) by using the relation $\nu = (2\pi c)^{-1} k^{1/2} [m^{-1} + M^{-1}]^{1/2}$. For undissociated stearic acid, $\Delta k = 4.8 \times 10^2 N/m$, for the nanocrystal COO bonds, $\Delta k = 5.9 \times 10^2 N/m$ whereas for the dissociated COO groups in all films, $\Delta k = \sim 2 \times 10^2 N/m$.

The values of k and Δk corresponding to the nanocrystal peaks are thus close to but distinctly different from those of undissociated acid, suggesting that the electron distributions in the two types of carbon-oxygen bond have been perturbed by possibly some supra-molecular bonding such as hydrogen bonding between the headgroups. This is consistent with recent findings of an interheadgroup hydrogen bonding as a precursor of monolayer collapse into nanocrystalline islands on water surface [?]. We therefore propose that these peaks arise due to the hydrogen bonded carboxylate headgroups at the nano-grain surfaces and interfaces.

9.1.4 Structure of Zinc Stearate Nanocrystals

In order to determine the exact molecular structure of the self-assembled ZnSt, X-Ray diffraction (XRD) measurements near the C K edge was carried out, using synchrotron radiation at BEAR beamline, Elettra Synchrotron. The results are discussed below. X-Ray diffraction results (Fig. 9.2) show that the ZnSt has self-assembled to form molecular crystals on both templates but have

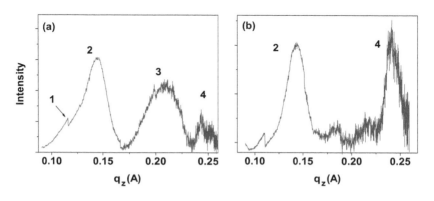

Figure 9.2: *X-ray diffraction near C k-edge (260 eV) of ZnSt nanocrystals on (a) Cd-T and (b) Co-T.*

Table 9.2: Results of XRD measurements

Peaks	Zn on Cd-T	Zn on Co-T	Reflection plane
1	51.93 (M)	-	(1/2,1/2,1/2)
2	43.33 (S)	43.94 (S)	(001)
3	30.50 (S)	-	(101)+(011)
4	26.65 (W)	25.96 (S)	(111)

S: Strong; M: Medium; W: Weak

difference in their structures. There are four diffraction peaks corresponding to ZnSt on Cd-T (Fig. 9.2(a)) and two peaks corresponding to ZnSt on Co-T (Fig. 9.2(b)). The prominent diffraction peaks along with their corresponding lattice planes are shown in Table 9.2. Peak 2, which appears to be equally intense in both cases (d $\sim$ 43 Å) correspond to headgroup separation in multilayers of untilted ZnSt molecules. Assuming d to be the lattice spacing c, peak 2 (d $\sim$ 43 Å) correspond to reflection from (001) planes. Accordingly, peak 3 (d $\sim$ 31 Å), which appears as a strong peak in ZnSt on Cd-T correspond to reflection from the planes (101) and (011). Peak 4 (d $\sim$ 26 Å), correspond to reflection from (111) plane. Intensity of this (111) peak is considerable for ZnSt on Co-T but very weak in case of Cd-T, consistent with a close packed structure for nanocrystals on Co-T and a non-close packed structure of the same on Cd-T. These structures for the nanocrystals are not observed in bulk ZnSt crystals [198]. Peak 1 (d $\sim$ 52 Å) is present in ZnSt on Cd-T only and has medium intensity. It probably corresponds to reflection from a (1/2, 1/2, 1/2) plane, i.e., some superlattice. Exact nature of this superlattice has not been ascertained but most probably it is coming from the fact that Zn-bearing carboxylate group has two structures unidentate and bidentate bridged. Absence of this superlattice in Co-T suggests a mixture of the two structures in each unit cell, as seen in ZnSt LB multilayers. Thus it is found that the in-plane forces present in parent templates cause a structural modification of the metal–organic composite self-assembled on it.

The behavior of ZnSt on MOTs to form self-assembled nanocrystals is in sharp contrast to its behavior on hydrphillic amorphous substrates on which it forms monomolecular layers ($\sim$ 25 Å). As a plausible explanation of this contrasting behavior, we would like to state that ZnSt wets the amorphous substrate in forming molecular layers having a contact angle close to zero, whereas it only critically wets the MOTs such that the contact angle is close to (but still less than) 90°. This accounts for the greater height of the ZnSt nanocrystals ($\sim$ 250 Å) obtained during monomolecular deposition. It must also be mentioned that the mechanism is not that of complete de-wetting since then the nanocrystals would not at all stay on the MOT surface during vertical deposition. Although further studies are required to establish this proposition, nevertheless this provides a novel route of obtaining ZnSt nanocrystals.

9.2 On Nanocrystal Growth in Langmuir–Blodgett Templates

Cadmuim stearate and cobalt stearate trilayers act as suitable templates and allow self-assembly of ZnSt nanocrystals. Combined studies by AFM and XRD near the C K edge show that the ZnSt nanocrystals have different structures when deposited on CdSt and CoSt templates, the former having non-close packed structure with superlattice ordering and the latter having a close packed structure without superlattice ordering. Presence (absence) of dipolar forces in the CdSt (CoSt) template is proposed to bring about this structural modification of ZnSt nanocrystals through a possible critical wetting mechanism.

Chapter 10

Conclusions and Outlook

10.1 What We Found

Langmuir–Blodgett film structure is known to depend strongly on the external deposition parameters like surface pressure, temperature, pH of the aqueous subphase etc. LB film research has been predominantly focused on controlling these experimental parameters to obtain perfectly periodic LB films. Compared to that, very few reports exist on plausible methods of manipulating the structure and bonding at the molecular level to tune film morphology at the micrometer scale. The work described in this book is an effort to relate the morphology and growth of such metal–organic multilayers to some aspects of their molecular structure, and to observe the effect of a change in the latter on the former.

The main results of our work may be summarized thus:

(1) There is a strong dependence of film morphology on molecular structure, and air–water interface plays a crucial role in mediating this dependence.

(2) The high free energy and inherent asymmetry of forces existing at the air–water interface leads to the co-existence of different metal–carboxylate bonding, that are not available in bulk.

(3) Air–water interface imparts flexibility to the metal-carboxylate system in selecting specific low energy conformations of these structures, different from those observed in bulk, by allowing formation of supramolecular bonds. These weak supramolecu-

lar bonds, due to their large number, are strong enough to build up near-perfect metal–organic multilayers depending upon the metal–organic interface. This result is of fundamental importance and shows that it is the interplay of these interactions at the molecular level that brings about a change in the morphology and that the above mentioned external parameters are only different ways of mediating such interactions.

(4) The metal–carboxylate bonds differ from each other by a crucial parameter, the net dipole moment, that can give rise to the presence or absence of long-range forces. Preference for these bonds or coordinations are decided by the atomic number of the metal concerned.

(5) These long-range forces decide the in-plane behavior of the film as "solid"-like or "liquid"-like.

The focus of our work was a basic understanding of how specific bonding determine the microscopic structure of these metal-bearing amphiphiles and, in turn, tunes the morphology of the films. Our primary finding was the dependence of film morphology on headgroup coordination in metal-bearing amphiphiles. From a study of Langmuir monolayers and Langmuir–Blodgett monolayers of cadmium stearate and cobalt stearate, we have shown that metal–carboxylate coordination is strongly dependent on the metal ion type, and that the LB transfer process mediates the specific selection quite elegantly. Our results showed that cadmium ions prefer unidentate coordination whereas cobalt ions prefer bidentate coordination both on water surface as well as on substrate surface. The difference is that in the former case all coordinations are found to co-exist, whereas in the latter case preferred coordinations were selected during transfer. Moreover, the transferred multilayers of the same metal stearates showed same coordination preference as its precursor monolayer.

A study of the morphology of these multilayers showed excellent out-of-plane crystalline growth in both cases with large reduction of in-plane defects in Co-bearing multilayers compared to Cd-bearing ones, indicating that bidentate bridged coordination plays a key role in stacking of defect-free multilayers.

We have investigated the sufficiency of the bidentate bridged coordination in forming LB multilayers. In doing so, we have also demonstrated the importance of air–water interface in selecting bond coordination and conformation on one hand and related the morphology to molecular conformations on the other. From a comparative study of cobalt stearate prepared in bulk and at air/water interface, both of which have identical metal ion-carboxylate coordination, viz. the bidentate bridged configuration, we have shown that defect-free multilayer formation takes place only in the latter case. We have associated this difference in their morphology with different conformations of their Co-bearing headgroups. Our results showed that preformed (bulk) cobalt stearate existed as a "boat" conformer with linear O-Co-O linkage, whereas the ones formed at air–water interface was a lower energy "twisted" conformer with a bent O-Co-O configuration.

We have shown that the flexible headgroup structure in the interfacial cobalt stearate allowed weak supramolecular interactions, of hyperconjugative origin, between methyl tops at chain-ends. These weak interactions coupled the hydrocarbon chains end-to-end and ordered their skeletal planes, thereby reducing free energy and leading to excellent Langmuir–Blodgett multilayers. On the other hand, the headgroup structure in the bulk prepared sample was very rigid and inhibited supramolecular bonding. As a result, multilayer growth could not take place.

We have studied the morphological evolution of different metal-bearing amphiphiles and correlated them with their individual headgroup configurations. Our results showed that for at least four transition metal stearates, viz. manganese stearate, cobalt stearate, zinc stearate and cadmium stearate, bidentate coordination of metal–carboxylate headgroup gave rise to layer-by-layer growth as observed in Mn, Co and partially in Zn. Crossover to island-like growth, as observed in Cd and Zn, was due to presence of unidentate coordination in headgroup. Each metal stearate film exhibited different growth modes, with Mn and Co bearing films showing *Frank Van der Merwe* type growth, Zn bearing films showing *Stranski–Krastanow* type growth and Cd bearing ones showing *Volmer Weber* type growth. We have shown that this growth mode varied with number (n) of metal atoms per headgroup, viz. $n = 1$ (in manganese and mobalt stearates), $n = 0.75$ (in zinc stearate) and $n = 0.5$ (in cadmium stearate), which in turn decided headgroup coordination such that $n = 1.0$ corresponds to bidentate and $n = 0.5$ corresponds to unidentate coordination; intermediate value in Zn corresponding to a mixture of these coordinations.

We have also shown that in bidentate headgroups, in-plane dipole moment was zero, which causes intermolecular forces between adjacent molecules to be absent allowing growth to proceed via layering. On the other hand, in unidentate headgroups, existence of non-zero in-plane dipole moment resulted in development of weak in-plane intermolecular forces between adjacent molecules causing island-like growth. Gradual decrease in n, in going from Mn to Cd, suggested that crossover from layering to island formation is not an abrupt process but rather systematic with change in atomic number of metal. This dependence on atomic number probably stems from the dependence of coordination on ionic radius of metal, the latter decreasing with decrease in atomic number.

The presence and absence of in-plane dipolar forces in unidentate and bidentate coordinated metal stearates point to a probable solid-like and liquid-like behavior, respectively. In order to ascertain this proposition, we have investigated the in-plane structural differences between the two amphiphilic salts cadmuim stearate and cobalt stearate by comparing the height correlations of their monolayers and multilayers. Our results showed that the Cd bearing films, with huge number of in-plane "pinhole" defects, exhibited self-affine behavior over a considerable scan range. On the other hand, the Co bearing films, which are almost devoid of such in-plane defects, do not strictly follow the self-affine behavior. These films show deviation from self-affinity for small scan sizes, suggesting more of a liquid-like behavior and flexible in-plane structure, the effect being more pronounced for the tri-layered film. Further, we have shown that cobalt stearate films exhibit an undulating topography which is probably achieved by the random orientation of the small tilts of adjacent hydrocarbon tails in a radially symmetric manner, suggesting that in CoSt, tail orientations help fill up in-plane defects and give rise to excellent in-plane coverage.

Finally, we have shown the potential of these *"solid-like"* and *"liquid-like"* films to behave as suitable templates for self-assembling nanocrystals of similar metal-bearing amphiphiles. More specifically, we have shown that cadmuim stearate and cobalt stearate trilayers can act as suitable templates that allow self-assembly of ZnSt nanocrystals. Preliminary results of x-ray diffraction show that the ZnSt nanocrystals have different structures when deposited on Cd-bearing and Co-bearing templates, the former having non-close packed structure with superlattice ordering and the latter having a close packed structure without superlattice ordering. Presence (absence) of dipolar forces in the Cd-bearing (Co-bearing) template is proposed to bring about this structural modification of ZnSt nanocrystals through a possible critical wetting mechanism.

10.2   What Lies Ahead

Plenty. Just one example. Our book involves the study of metal–carboxylate interactions both on water surface and on substrate surface. The effect of these interactions on morphology has been studied for films on the substrate surface. However, we did not "look at" the surface of the corresponding Langmuir monolayers. From the results discussed so far, one would expect the morphology of the precursor monolayer to be modified in presence of different metal ions. More specifically, it is worthwhile to investigate whether Cd and Co ions behave identically on the water surface as they do on the amorphous substrate surface. Results would suggest whether the distinct behavior of CdSt and CoSt films are inherent, thereby bringing out the significance of substrate in mediating the metal–carboxylate interactions. Some work has been carried out in this direction, preliminary result of which are discussed here.

We have carried out Brewster Angle Microscopy (BAM) of stearic acid monolayers (Fig. 10.1) with Cd and Co ions in

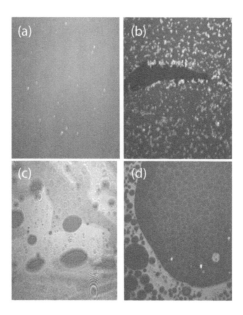

Figure 10.1: *BAM images of stearic acid Langmuir monolayer with (a), (b) Cd and (c), (d) Co ions in subphase. The images are taken during decompression of the isotherm at (a), (c): 30 mN/m and (b), (d) 5 mN/m.*

subphase. Preliminary results show that the monolayers behave distinctly in presence of these ions. The images correspond to monolayer decompression for two values of surface pressure viz. $\pi_s = 30$mN/m and $\pi_s = 5$mN/m. With cadmium ions in subphase, the compressed stearic acid monolayer (Fig. 10.1(a)) shows the formation of "crystallites" that remain unaffected when decompressed (Fig. 10.1(b)). The CdSt film shows the presence of a monolayer on which these crystallites are formed. This is clear indication of CdSt multilayer formation at air–water interface and is consistent with our result (discussed in Chapter 7), which shows that for CdSt the 1ML film is not a true monolayer but is deposited as a multilayer. Further, presence of some "cracks" in this underlying monolayer is a clear indication of the fact that the CdSt system is a rigid one.

In presence of cobalt ions, on the other hand, the monolayer "spreads out" gradually as π_s decreases (Fig. 10.1(c) and (d)) and forms completely interconnected "soap bubble-like" features on decompression. These aggregate continuously to form the monolayer on re-compression. These decompression and re-compression behavior of Langmuir monolayers with cadmium and cobalt ions in subphase suggest, respectively, the essentially irreversible fracture of a "solid" and the reversible, interconnected-bubble features of a soapy "liquid." It is thus evident that the CoSt system is inherently a more flexible system both at the air–water interface and the air-substrate interface and its "liquid-like" behavior arises probably due to the specific interaction of metal ion with the organic headgroup and not due to the transfer process.

It is to be noted that the connection between this interaction and the "liquid" or "solid" like features has been discussed for Langmuir–Blodgett multilayers, in Chapters 7 and 8 of this book. However, it has not been clarified for Langmuir monolayers and needs to be done.

Again, this work may be further extended to study the behavior of fatty acid Langmuir Monolayers with other mono and divalent metal ions in subphase. A rather interesting study would involve the effect of surface pressure and change in pH on the precursor monolayer surface as well as on metal–carboxylate interactions. We have studied the effect of barrier compression on metal–carboxylate coordination in stearic acid Langmuir monolayers containing Cd ions in subphase. Preliminary results of IRAS indicate that with increase in surface pressure, the Cd-carboxylate coordination shows strong preference towards asymmetric headgroup structure, i.e., the unidentate coordination. Although the results are at a nascent stage, they strongly indicate that increase in surface pressure at the water surface and the LB transfer process has identical effect on the metal–carboxylate coordination. This is consistent with previous work [206]. In general, it is worthwhile to conduct a systematic study of the dependence of metal–carboxylate coordinations on the external parameters especially, pH, temperature and compression, rate that control the morphology of the metal–organic films.

However, the aforementioned studies focus on specific extensions of our research. Two of our findings may have more general and extensive interest. First, the definite role of the air–water interface in selecting bond coordination and conformation, which is probably connected to the observation of hydronium ion enrichment at water surface [193]. Second, correlating morphology to bonding molecules via molecular conformations suggest that specific supramolecular forces and large amplitude motions have a significant part to play. This is because these are allowed or inhibited by molecular conformations. These results may be utilized to tune self-organization of more complex structures pertaining, for example, to biomimetic systems, by the low-energy modes of large amplitude motions.

Monomolecular layers of amphphiles provide the "simplest" example of complex two-dimensional systems and that is the key of its continuing attraction over more than a century of research. The physics of pristine monolayers have been studied over this long period and though the phases and their transitions have been "understood" in the framework of Landau theory yet it is clear that such understanding is incomplete since the monolayers are in a state of metastable equilibrium. Monolayer dynamics, especially long term dynamics, has not been simulated successfully. Nor do we understand fully the dynamics of transfer from liquid to solid surfaces.

If this is the state of affairs with pristine monolayers, it is easy to imagine the scope of research with mixed monolayers or, as in our systems, monolayers reacting with different chemical species. In our first hesitant steps into this vast unknown territory we chose one of the simplest reactions - formation of amphiphilic fatty acid salts and came into the unexpected richness presented in this book. We rest our case.

Bibliography

[1] G. Roberts, *Langmuir–Blodgett Films* (Plenum Press, New York, 1990).

[2] I. R. Peterson, *J. Phys. D: Appl. Phys* **23**, 379 (1990).

[3] A. Ulman, *Introduction to Ultrathin Organic Films* (Academic, New York, 1991).

[4] R. H. Tredgold, *Order in Organic Thin Films* (Cambridge University Press, Cambridge, 1994).

[5] M. C. Petty, *Langmuir–Blodgett Films: An Introduction* (Cambridge University Press, Cambridge, 1996).

[6] A. Riul Jr., H. C. de Souza, R. R. Malmegrim, D. S. dos Santos Jr., A. C. P. L. F. Carvalho, F. J. Fonseca, O. N. Oliveira Jr., and L. H. C. Mattoso, *Sens. Act. B* **98**, 77 (2004).

[7] C. E. Borato, A. Riul Jr., M. Ferreira, O. N. Oliveira Jr., and L. H. C. Mattoso, *Instrum. Sci. and Technol.* **32**, 21 (2004).

[8] M. Ferreira, C. J. L. Constantino, A. Riul Jr., K. Wohnrath, R. F. Aroca, J. A. Glacometti, O. N. Oliveira Jr., and L. H. C. Mattoso, *Polymer* **44**, 4205 (2003).

[9] D. T. Balogh, C. R. Mendoncaa, A. Dhanabalan, S. Major, S. S. Talwar, S. C. Zilio, and O. N. Oliveira Jr., *Material Chem. Phys.* **80**, 541 (2003).

[10] C. R. Mendoncaa, D.S. dos Santos Jr., D. T. Balogh, G. A. Giaconetti, S.C. Zilio, and O.N. Oliveira Jr., *Polymer* **42**, 6539 (2001).

[11] L. M. Bilinov, S. P. Palto, and S. G. Yudin, *Appl. Phys. Lett.* **80**, 16 (2002).

[12] Y. Zong, K. Tawa, B. Menges, J. Ruhe, and W. Knoll, *Langmuir* **21**, 7036 (2005).

[13] L. M. Bilinov, V. M. Fridkin, S. P. Palto, A. V. Sorokin, and S. G. Yudin, *Thin Solid Films* **284**, 474 (1996).

[14] A. Bune, S. Ducharme, L. M. Bilinov, V. M. Fridkin, S. P. Palto, N. Petukhova, and S. G. Yudin, *Appl. Phys. Lett.* **67**, 3975 (1995).

[15] F. de Moura, and M. Trisic, *J. Phys. Chem B.***109**, 4032 (2005).

[16] Z. Biana, K. Wanga, L. Jina, and C. Huangh, *Coll. Surf. A: Physicochemical and Engineering Aspects* **257**, 67 (2005).

[17] K. Han, Q. Wang, G. Tang, H. Li, X. Sheng, and Z. Huang, *Thin Solid Films* **476**, 152 (2005).

[18] K. M. Mayya, N. Jain, A. Gole, D. Langevin, and M. Sastry, *J. Coll. Int. Sc.* **270**, 133 (2004).

[19] D. Natalia, M. sanpietro, L. Franco, A. Bolognesi, and C. Bottac, *Thin Solid Films*, **472**, 238 (2005).

[20] G. G. Roberts, M. C. Petty, S. Baker, M. T. Fowlers, and N. J. Thomas, *Thin Solid Films* **132**, 113 (1985).

[21] T. J. Rece, S. Ducharme, A. V. Sorokin, M. Poulsen, *Appl. Phys. Lett.* **82**, 142 (2003).

[22] N. Tancrez, C. Feuvrie, I. Ledoux, J. Zyss, L. Toupet, H. L. Bozec, and D. Maury, *J. Am. Chem. Soc.* **127**, 13474 (2005).

[23] T. Verbiest, S. Sioncke, and G. Koeckelberghs, *Chem. Phys. Lett.* **404**, 112 (2005).

[24] J. W. Boldwin, R. R. Amaresh, I. R. Peterson, J. W. Shumate, P. M. Cava, A. M. Amiri, R. Hamilton, G. Ashwell, and R. M. Metzger, *J. Phys. Chem. B* **106**, 12158 (2002).

[25] I. Langmuir, *J. Am. Chem. Soc.* **39**, 1848 (1917).

[26] G. L. Gaines, *Insoluble Monolayers at Liquid-Gas Interfaces* (Interscience, New York, 1966).

[27] V. M. Kaganer, H. Möhwald, and P. Dutta, *Rev. Mod. Phys.* **71**, 779 (1999).

[28] M. Yazdanian, H. Hu, and G. Zografi, *Langmuir* **6**, 1093 (1990).

[29] S. S. Stenhagen, and E. Stenhagen, *Nature* **156**, 239 (1945).

[30] M. Shih, T. Bohanon, J. Mikrut, P. Zschack, and P. Dutta, *J. Chem Phys.* **96**, 1556 (1992).

[31] C. Böhm, F. Leveiller, D. Jacquemain, H. Möhwald, K. Kjaer, J. Als-Nielsen, I. Weissbuch, and L. Leiserowitz, *Langmuir* **10**, 830 (1994).

[32] A. Datta, J. Kmetko, A. G. Richter, C.-J. Yu, and P. Dutta, *Langmuir* **16**, 1239 (2000).

[33] C. Ybert, W. Lu, G. Möller, and C. M. Knobler, *J. Phys. Chem. B* **106**, 2004 (2002).

[34] Y.-L. Lee, and K.-L. Liu, *Langmuir* **20**, 3180 (2004).

[35] V. M. Kaganer, and E. B. Loginov, *Phys. Rev. Lett.* **71**, 2599 (1993).

[36] V. M. Kaganer, and E. B. Loginov, *Phys. Rev. E* **51**, 2237 (1995).

[37] K. B. Blodgett, *J. Amer. Chem. Soc.* **57**, 1007 (1935).

[38] K. B. Blodgett, and I. Langmuir, *Phys. Rev.* **51**, 964 (1937).

[39] S. Mann, D. D. Archibald, J. M. Didymus, T. Douglas, B. R. Heywood, F. C. Meldrum, and N. J. Reeves, *Science* **261**, 1286 (1993).

[40] J. Kmetko, C. Yu, G. Evmenenko, S. Kewalramani, and P. Dutta, *Phys. Rev. Lett.* **89**, 186102 (2002).

[41] A. Datta, J. Kmetko, C.-J. Yu, A. G. Richter, K.-S. Chung, J.-M. Bai, and P. Dutta, *J. Phys. Chem. B* **104**, 5797 (2000).

[42] F. Leveiller, D. Jacquemain, M. Lahav, L. Leiserowitz, M. Deutsch, K. Kjaer, and J. Als-Nielsen, *Science* **252**, 1532 (1991).

[43] F. Leveiller, C. Böhm, D. Jacquemain, M. Lahav, H. Möhwald, L. Leiserowitz, K. Kjaer, and J. Als-Nielsen, *Langmuir* **10**, 819 (1994).

[44] J. Kmetko, A. Datta, G. Evmenenko, M. K. Durbin, A. G. Richter, and P. Dutta, *Langmuir* **17**, 4697 (2001).

[45] J. Kmetko, A. Datta, G. Evmenenko, and P. Dutta, *J. Phys. Chem. B* **105**, 10818 (2001).

[46] Y. Ren, K.-I. Iimura, and T. Kato, *Langmuir* **17**, 2688 (2001).

[47] D. K. Schwartz, *Surf. Sci. Rep.* **27**, 241 (1997).

[48] R. Ghaskadvi, S. Carr, and M. Denim, *J. Chem. Phys.* **111**, 3675 (1999).

[49] J. Zasadzinski, R. Viswanathan, L. Madsen, J. Garnaes, and D. K. Schwartz, *Science* **263**, 1726 (1994).

[50] J. K. Basu, and M. K. Sanyal, *Phys. Rep.* **363**, 1 (2002).

[51] J. B. Peng, G.T. Barnes, and I. R. Gentle, *Adv. Coll. Interface Sc.* **91**, 163 (2001).

[52] M. K. Mukhopadhyay, M. K. Sanyal, A. Datta, M. Mukherjee, T. Geue, J. Grenzer,and U. Pietsch, *Phys. Rev. B* **70**, 245408 (2004).

[53] E. Hatta, T. Maekawa, K. Mukasa, and Y. Shimoyama, *Phys. Rev. B* **60**, 14561 (1999).

[54] M. K.Mukhopadhyay, M. K. Sanyal, M. D. Mukadam, S. M. Yusuf, and J. K. Basu, *Phys. Rev. B* **68**, 174427 (2003).

[55] P. Facci, A. Diaspro, and R. Rolandi, *Thin Solid Films* **327**, 532 (1998).

[56] A. G. Milekhin, L. L. Sveshnikova, S. M. Repinsky, A. K. Gutakovsky, M. Friedrich, and D.R.T. Zahn, *Thin Solid Films* **422**, 200 (2002).

[57] J. K. Basu, and M. K. Sanyal, *Phys. Rev. Lett.* **79**, 4617 (1997).

[58] R. C. Ehlert, *J. Coll. Sci.* **20**, 387 (1965).

[59] A. Malik, M. K. Durbin, A. G. Richter, K. G. Hung, and P. Dutta, *Phys. Rev. B* **52**, R11654 (1995).

[60] U. Englisch, F. Peñacorda, I. Samoilenko, U. Pietsch, *Physica B* **248**, 258 (1998).

[61] A. I. Kitaigorodsky, *Organic Chemical Crystallography* (Consultants Bureau, New York, 1961).

[62] M. K. Sanyal, M. K. Mukhopadhyay, M. Mukherjee, A. Datta, J. K. Basu, and J. Penfold, *Phys. Rev. B* **65**, 033409 (2002).

[63] S. Kundu, A. Datta, and S. Hazra, *Langmuir* **21**, 5894 (2005).

[64] S. Kundu, *Coll. Surf. A* **348**, 196 (2009).

[65] A. Datta, S. Kundu, M. K. Sanyal, J. Daillant, D. Luzet, C. Blot, and B. Struth, *Phys. Rev. E* **71**, 041604 (2005).

[66] M. V. Frieling, and H. Bradaczek, *Acta Cryst. A* **46**, 227 (1990).

[67] Y. Sasanuma, and H. Nakahara, *Thin Solid Films* **261**, 280 (1995).

[68] D. K. Schwartz, R. Viswanathan, and J. A. Zasadzinski, *J. Am. Chem. Soc.* **115**, 7374 (1993).

[69] Th. Geue, M. Schultz, U. Englisch, R. Stommer, U. Pietsch, K. Meine, and D. Vollhardt, *J. Chem. Phys.* **110**, 8104 (1999).

[70] X.-H. Li, M. Li, and Z.-H. Mai, *Phys. Rev. B* **69**, 235407 (2004).

[71] D. Y.Takamoto, E. Aydil, J. A. Zasadzinski, A. T. Ivanova, D. K. Schwartz, T. Yang, and P. S. Cremer, *Science* **293**, 1292 (2001).

[72] A. Gibaud, N. Cowlam, G. Vignaud, and T. Richardson, *Phys. Rev. Lett.* **74**, 3205 (1995).

[73] J. K. Basu, S. Hazra, and M. K. Sanyal, *Phys. Rev. Lett.* **82**, 4675 (1999).

[74] R. Viswanathan, J. A. Zasadzinski, and D. K. Schwartz, *Science* **261**, 449 (1993).

[75] K. Ueno, K. Saiki, T. Shimada, A. Koma, *J. Vac. Sci. Tech. A* **8**, 68 (1990).

[76] A. Koma, K. Saiki, Y. Sato, Appl. *Surf. Sci.* **41**, 451 (1989).

[77] F. S. Ohuchi, B. A. Parkinson, K. Ueno, A. Koma, *J. Appl. Phys.* **68**, 2168 (1990).

[78] C. J. Eckhardt et al., *Langmuir* **8**, 2591 (1992).

[79] C. J. Eckhardt and D. R. Swanson, *Chem. Phys. Lett.* **194**, 370 (1992).

[80] R. Krechetnikov and G. M. Homsy, *J. Fluid Mech.* **559**, 429 (2006).

[81] I. Daruka, and A. L. Barabási, *Phys. Rev. Lett.* **79**, 3708 (1997).

[82] J. H. Van der Merwe, *Interface Sci.* **1**, 1 (1993).

[83] F. C. Frank, and J. H. van der Merwe, *Proc. Roy. Soc. Ser. A* **198**, 205 (1949).

[84] J. W. Matthews, *Epitaxial Growth*, Part 2, Ed. J.W. Matthews (Academic Press, New York, 1975).

[85] J. H. van der Merwe, *J. Appl. Phys.* **34**, 123 (1963).

[86] Gendry, V. Drouot, C. Santinelli, G. Hollinger, C. Miossi, and M. Pitaval, *J. Vat. Sci. Technol. B* **10**, 1829 (1992).

[87] P. R. Berger, K. Chang, P. Bhattacharya, J. Singh, and K. K. Bajaj, *Appl. Phys. Lett.* **53**, 684 (1988).

[88] E. Bauer, *Z. Kristallogr.* **110**, 423 (1958).

[89] L. Pauling, *The Nature of the Chemical Bond and the Structure of Molecules and Crystals: An Introduction to Modern Structural Chemistry* (Cornell University Press, U.S.A., 1960).

[90] K. Nakamoto, *Infrared Spectra of Inorganic and Coordination Compounds* (John Wiley & Sons, New York, 1963).

[91] J. O. Bockris, and A. K. N. Reddy, *Modern Electrochemistry I: Ionics* (Plenum Press, New York, 1998).

[92] N. P. Kumar, S. S. Major, S. Vitta, S. S. Talwar, A. Gupta, and B. A. Dasannacharya, *Colloid Surface A* **257**, 243 (2005).

[93] A. Gericke, and H. Huhnerfuss, *Thin Solid Films* **245**, 74 (1994).

[94] H.-D. Lin, and D.-C. Ho, *Int. J. Adv. Manuf. Technol.* **34**, 567 (2007).

[95] A. Whitty, *Nature Chem. Biol.* **4**, 435 (2008).

[96] E. Sharfrin, and W. A. Zisman, *J. Phys. Chem.* **64**, 519 (1960).

[97] N. Eustathopoulos, M. G. Nicholas, and B. Drevet, *Wettability at High Temperatures* (Pergamon, Oxford, UK, 1999).

[98] M. E. Schrader, and G. I. Loeb, *Modern Approaches to Wettability: Theory and Applications.* (Plenum Press, New York, 1992).

[99] P. G. de Gennes, *Rev. Mod. Phys.* **57**, 827 (1985).

[100] R. E. Johnson, *Wettability* Ed. John. C. Berg, (Marcel Dekker, Inc., New York, 1993)

[101] K. S. Lee, N. Ivanova, V. M. Starov, N. Hilal, and V. Dutschk, *Adv. Coll. Interface Sci.* **144**, 54 (2008).

[102] P. G. de Gennes, *Soft Interfaces* (Cambridge University Press, UK, 1997).

[103] L. Wilhelmy, *Ann. Phy.* **119**, 177 (1863).

[104] A. J. G. Allan, *J. Coll. Sci.* **13**, 273 (1958).

[105] W. D. Harkins, and T. F. Anderson, *J. Am. Chem. Soc.* **59**, 2189 (1937).

[106] D. G. Dervichian, *Ann. Phys. 11* **8**, 361 (1937).

[107] W. Rabinovitch, R. F. Robertson, and S. G. Mason, *Can. J. Chem.* **38**, 1881 (1960).

[108] F. H. C. Stewart, *Australian J. Chem.* **14**, 159 (1961).

[109] S. Kundu, A. Datta, and S. Hazra, *Langmuir* **24**, 9386 (2008).

[110] I. Langmuir, and V. J. Schaefer, *J. Am. Chem. Soc.* **60**, 1351 (1938).

[111] T. Kato, *Jpn. J. Appl. Phys.* **27**, L2128 (1988).

[112] K. Y. C. Lee, M. M. Lipp, D.Y. Takamoto, E. Ter-Ovanesyan, A. Zasadzinski, and A. J. Waring, *Langmuir* **14**, 2567 (1998).

[113] T. Kato, *Jpn. J. Appl. Phys.* **26**, L1377 (1987).

[114] T. Kato, N. Matsumoto, M. Kawano, N. Suzuki, T. Araki, and K. Iriyama, *Thin Solid Films* **242**, 223 (1994).

[115] R. Wiesendanger, *Scanning Probe Microscopy and Spectroscopy* (Cambridge University Press, Cambridge, 1994).

[116] D. Sarid, *Scanning Force Microscopy* (Oxford University Press, New York, 1994).

[117] C. J. Chen, *Introduction to Scanning Tunneling Microscopy* (Oxford University Press, New York, 1993).

[118] A. Noy, C. H. Sanders, D. V. Vezenov, S. S. Wong, and C. M. Lieber, *Langmuir* **14**, 1508 (1998).

[119] S. N. Magonov, V. Elings, and M. H. Whangbo, *Surf. Sci.* **375**, L385 (1997).

[120] G. Bar, Y. Thomann, R. Brandsch, H. J. Cantow, and M. H. Whangbo, *Langmuir* **13**, 3807 (1997).

[121] N. A. Burnham et al., *Nanotechnology* **8**, 67 (1997).

[122] M. Tsukuda, N. Sasaki, R. Yamura, N. Sato, and K. Abe, *Surf. Sci.* **401**, 355 (1998).

[123] J. P. Spatz, S. Sheiko, M. Möller, R. G. Winkler, P. Reineker, and O. Marti, *Langmuir* **13**, 4699 (1997).

[124] B. Anczykowski, D. Krüger, K. L. Babcock, and H. Fuchs, *Ultramicroscopy* **66**, 251 (1996).

[125] T. E. Schäffer, J. P. Cleveland, F. Ohnesorge, D. A. Walters, and P. K. Hansma, *J. Appl. Phys.* **80**, 3622 (1996).

[126] T. P. Russell, *Mat. Sci. Rep.* **5**, 171 (1990).

[127] J. Als-Nielsen, D. Jacquemain, K. Kjaer, F. Leveiller, M. Lahav, and L. Leiserowitz, *Phys. Rep.* **246**, 251 (1994).

[128] J. Als-Nielsen and D. McMorrow, *Elements of Modern X-Ray Physics* (John Wiley & Sons, New York, 2000).

[129] X-Rays 100 Years Later, *Phys. Today (Special Issue)* **48**, (1995).

[130] C.-J. Yu, A. G. Richter, A. Datta, M. K. Durbin and, P. Dutta, *Phys. Rev. Lett.* **82**, 2326 (1999).

[131] A. Braslau, P. S. Pershan, G. Swislow, B. M. Ocko, and J. Als-Nielsen, *Phys. Rev. A* **38**, 2457 (1986).

[132] M. K. Sanyal, S. K. Sinha, K. G. Huang, and B. M. Ocko, *Phys. Rev. Lett.* **66**, 628 (1991).

[133] J. Als-Nielsen, *Handbook of Synchrotron Radiation* (Eds. G. S. Brown and D. E. Moncton) **3**, 471 (1991).

[134] M. Tolan, *X-Ray Scattering from Soft Matter Thin Films, Springer Tracts in Modern Physics*, vol 148 (Springer, Berlin, 1999).

[135] J. Daillant, and A. Gibaud, *X-Ray and Neutron Reflectivity: Principles and Applications* (Springer, New York, 1999).

[136] V. Hóly, U. Pietsch, and T. Baumbach, *X-Ray Scattering from Thin Films and Multilayers, Springer Tracts in Modern Physics* Vol. 149 (Springer, Berlin, 1999).

[137] M. K. Sanyal, A. Datta, and S. Hazra, *Pure Appl. Chem.* **74**, 1553 (2002).

[138] S. K. Sinha, E. B. Sirota, S. Garoff, and H. B. Stanley, *Phys. Rev. B* **38**, 2297 (1988).

[139] G. H. Vineyard, *Phys. Rev. B* **26**, 4146 (1982).

[140] I. K. Robinson, and D. J. Tweet, *Rep. Prog. Phys.* **55**, 599 (1992).

[141] R. W. James, *The Optical Principles of the Diffraction of X-Rays* (Ox Bow Press, Woodbridge, Connecticut, 1982).

[142] *International Tables for Crystallography*, edited by A. J. C. Wilson (Kluwer Academic, Dordrecht, Boston, London, 1992).

[143] M. Born, and E. Wolf, *Principles of Optics*, (Pergamon Press, New York, 1985).

[144] D. J. Griffiths, *Introduction to Electrodynamics* (Prentice Hall, New Jersey, 1989)

[145] J.D. Jackson, *Classical Electrodynamics* (John Wiley and Sons, New York, 1998).

[146] Brian C. Smith, *Fundamentals of Fourier Transform Infrared Spectroscopy* (CRC Press, New York, 1996).

[147] J. Stohr, *NEXAFS Spectroscopy* (Springer-Verlag, New York, 1996).

[148] B. K. Teo, *EXAFS: Basic Principles and Data Analysis*, (Spinger - Verlag, New York, 1986).

[149] P. A. Lee, P. H. Citrin, P. Eisenberger, and B. M. Kincaid, *Rev. Mod. Phys.* **53**, 769(1981).

[150] S. Nannarone, F. Borgatti, A. DeLuisa, B. P. Doyle, G. C. Gazzadi, A. Giglia, P. Finetti, N. Mahne, L. Pasquali, M. Pedio, G. Selvaggi, G. Naletto, M.G. Pelizzo, and G. Tondello, *AIP Conf. Proc.* **705**, 450 (2004).

[151] G. Palasantzas, and J. Krim, *Phys. Rev. B* **48**, 2873 (1993).

[152] H. -N. Yang, Y. -P. Zhao, A. Chan, T. -M. Lu, and G. -C. Wang, *Phys. Rev. B* **56**, 4224 (1997).

[153] C. A. Croxton, *Statistical Mechanics of the Liquid Surface* (John Wiley & Sons, New York, 1980).

[154] J. Daillant, and M. Alba, *Rep. Prog. Phys.* **63**, 1725 (2000).

[155] J. Daillant, J. Pignat, S. Cantin, F. Perrot, S. Morac, and O. Konovalovd, *Soft Matter* **5**, 203 (2009).

[156] L. G. Parratt, *Phys. Rev.* **95**, 359 (1954).

[157] A. Gibaud, and S. Hazra, *Curr. Sci.* **78**, 1467 (2000).

[158] P. R. Bevington, *Data Reduction and Error Analysis for the Physical Sciences*, (Mcgraw-Hill, New York, 1969).

[159] L. D. Landau, and E. M. Lifshitz, *Mechanics*, 3rd. ed. (Pergamon Press, 1976).

[160] E. B. Wilson, J. C. Decius and P. C. Cross, *Molecular Vibrations* (McGraw-Hill, 1955; Reprinted by Dover 1980).

[161] K. Nakamoto, *Infrared and Raman Spectra of Inorganic and Coordination Compounds* (John Wiley & Sons, New York, 1970).

[162] H. Goldstein, *Classical Mechanics* (Addison-Wesley, U.S.A., 1980).

[163] N. B. Colthup, L. H. Daly, and S. E. Wiberly *Introduction to Infrared and Raman Spectroscopy* (Academic Press: New York, 1965).

[164] B. Schrader, *Infrared and Raman Spectroscopy: Methods and Applications* (VCH Publishers, NewYork, 1995).

[165] B. Qiao, J. J. Cerdà, and C. Holm, *Macromolecules* **44**, 1707 (2011).

[166] W. Kemp, *Organic Spectroscopy* (Palgrave, Hampshire, 2002).

[167] J. F. Rabolt, F. C. Burns, N. E. Schlotter, J. D. Swalen, *J. Chem. Phys.* **78**, 946 (1983).

[168] E. G. Palacios, G. Juarez-Lopez, A. J. Monhemius, *Hydrometallurgy* **72**, 139 (2004).

[169] Y. Lu and J. D. Miller, *J. Coll. Interface Sci.* **256**, 41 (2002).

[170] Chang Ok Kim et al., *Langmuir* **19**, 4504 (2003).

[171] S. G. Urquhart, A. P. Hitchcock, A. P. Smith, H. Ade, and E.G. Rightor, *J. Phys. Chem. B* **101**, 2267 (1997) .

[172] S. G. Urquhart, and H. Ade, *J. Phys. Chem. B* **106**, 8531 (2002).

[173] F. Kimura, J. Umemura, and T. Takenaka, *Langmuir* **2**, 96 (1986).

[174] N. Wu, L. Fu, M. Su, M. Aslam, K. C. Wong, and V. P. Dravid, *Nano Lett.* **4**, 383 (2004).

[175] J. A. Larabee, C. H. Leung, R. L. Moore, T. Thamrong-Nawasawat, and B. S .H. Wessler, *J. Amer. Chem. Soc.* **126**, 12316 (2004).

[176] A. Kremer-Aach, W. Kla1ui, R. Bell, A. Strerath, H. Wunderlich, and D. Mootz, *Inorg. Chem.* **36**, 1552 (1997).

[177] J. Kmetko, A. Datta, G. Evmenenko, and P. Dutta, *J. Phys. Chem. B*, **105**, 10818 (2001).

[178] C. Bollinne, S. Cuenot, B. Nysten, and A. M. Jonas, *Eur. Phys. J. E* **12**, 389 (2003).

[179] J. N. Israelachvili, *Intermolecular and Surface Forces* (Academic Press, London, 1991).

[180] S. R. Ahmed, and P. Kofinas, *J. Magn. Magn. Mater.*, **288**, 219 (2005).

[181] I. A. Nekrasov, S. V. Streltsov, M. A. Korotin, V. I. Anisimov, *Phys. Rev. B* **68**, 235113 (2003).

[182] D. G. Lister, *Internal Rotation and Inversion* (Academic Press, London, 1978).

[183] A. Allerhand, and P. V. R. Schleyer, *J. Am. Chem. Soc.* **85**, 1715 (1963).

[184] I. V. Alabugin, M. Manoharan, S. Peabody, and F. Weinhold, *J. Am. Chem. Soc.* **125**, 5973 (2003).

[185] I. V. Alabugin, *J. Org. Chem.* **65**, 3910 (2000).

[186] V. Pophristic, and L. Goodman, *Nature* **411**, 565 (2001).

[187] R. W. Corkery, and D. C. R.Hockless, *Acta Cryst.* **C53**, 840 (1997).

[188] R. F. Holland, and J. R. Nielsen, *J. Mol. Spectrosc.* **9**, 436460 (1962).

[189] D. A. Outka, J. Stohr, J. P. Rabe, and J. D. Swalen, *J. Chem. Phys.* **88**, 4076 (1987).

[190] P. S. Bagus, K. Weiss, A. Schertel, Ch. Wll, W. Braun, C. Hellwig, and C. Jung, *Chem. Phys. Lett.* **248**, 129, 1996.

[191] A. P. Hitchcock, and I. Ishii, *J. Electron Spectrosc. Relat. Phenom.* **42**, 11, 1987.

[192] S. Malcharek, A. Hinz, L. Hilterhaus, and H. J. Galla, *Biophys. J.* **88**, 2638, 2005.

[193] P. B. Petersen, and R. J. Saykally, *Chem. Phys. Lett.* **458**, 255, 2008.

[194] S. Kundu, A. Datta, and S. Hazra, *Chem. Phys. Lett.* **405**, 282 (2005).

[195] S. Kundu, A. Datta, M. K. Sanyal, J. Daillant, D. Luzet, C. Blot, and B. Struth, *Phys. Rev. E* **73**, 061602 (2006).

[196] M. Li, X. H. Li, L. Huang, Q. J. Jia, W. L. Zheng, and Z. H. Mai, *Europhys. Lett.* **64**, 385 (2003).

[197] G. Palasantzas, *Solid State Comm.* **100**, 705 (1996).

[198] T. Ishioka, K. Maeda, I. Watanabe, S. Kawauchi, M. Harada, *Spectrochim. Acta A* **56**, 1731 (2000).

[199] S. A. Darst, H. 0. Ribi, D. W. Pierce and R. D.Kornberg, *J. Mol. Biol.*, 1988, **203**, 269.

[200] H. 0. Ribi, D. S. Ludwig, K. L. Mercer, G. K. Schoolnik and R. D. Kornberg, *Science*, 1988, **239**, 1272.

[201] J. K. Pike, H. Byrd, A. A. Morrone and D. R. Talham, *J. Am. Chem. Soc.*, 1993, **115**, 8497.

[202] P. Müller and R. Kern, *Appl. Surf. Sci.*, 2000, **162**, 133.

[203] K. Georgsson, N. Carlsson, L. Samuelson, W. Seifert and L. R. Wallenberg, *Appl. Phys. Lett.*, 1995, **67**, 2981.

[204] Z. Zhu, A. Zhang, G. Ouyang and G. Yang, *J. Phys. Chem. C* , 2011, **115**, 6462.

[205] F. Boxberg and J. Tulkki, *Rep. Prog. Phys.*, 2007, **70**, 1425.

[206] R. S. Ghaskadvi, J. B. Ketterson, and P. Dutta, *Langmuir* **13**, 5137 (1997).

Index

201